2.00

REMARKABLE ANSWERS TO PROVOCATIVE QUESTIONS ABOUT LIFE . . . AND BEYOND!

THE STAIRWAY TO HEAVEN

"A dazzling performance . . .
Sitchin is a zealous investigator."
Kirkus Reviews

"Compelling."
Booklist

"Well-researched and persuasive . . .
Sitchin brings to this book the logic and
scholarship so often lacking . . .
He demonstrates the flaws
in establishing theories about pyramid-builders
and exposes a few ancient hoaxes."
Library Journal

"Impressive."
United Press International

"An intriguing book—read it . . .
Sitchin is to be congratulated . . .
He makes his point, and makes it well."
Kirkus Reviews

Avon Books by
Zecharia Sitchin

THE EARTH CHRONICLES

ZECHARIA
SITCHIN

THE STAIRWAY TO HEAVEN
BOOK II OF THE EARTH CHRONICLES

HARPER
An Imprint of HarperCollins*Publishers*

HARPER

An Imprint of HarperCollins*Publishers*
10 East 53rd Street
New York, New York 10022-5299

First Harper paperback printing: April 2007
First Avon Books paperback printing: May 1983

HarperCollins® and Harper® are trademarks of HarperCollins Publishers.

Printed in the United States of America

Visit Harper paperbacks on the World Wide Web at
www.harpercollins.com

10 9 8 7 6 5 4 3

Contents

Foreword

The Stairway to Heaven has been rightly considered a study in the origins of Man's unending search for immortality. It differs, however, from all others on the subject in that it treats ancient tales of such a search not as allegories, utopias, or fantasies, but rather as actual journeys to actual places on Earth.

The tale of Gilgamesh, the Sumerian king of Uruk (the biblical Erech), who sought the immortality to which he was entitled by being two-thirds divine, has been told before; but in this book it will be shown that he had actually gone to two (not just one) space-connected sites, whose existence will be proven and whose locations will be revealed. We will follow in the footsteps of Canaanite kings who thought they had escaped mortality by having been to the place of the gods' Fiery Stones—and will show where they went, what was there, and why the space-related sites have remained focal points in past events and present-day turmoil.

While the previous book, *The Twelfth Planet*, relied primarily on the Mesopotamian evidence, this volume adds the Egyptian dimension and shows that when Egyptian pharaohs departed for an Afterlife Journey to the Planet of Millions of Years, they embarked on a virtual journey to a place—an actual place—from which, outfitted as astronauts, they could be lifted to the planet of the gods.

Why were the pyramids of Giza called by them "Stairways to Heaven"? Who built them, why, and when? The amazing answers include a meticulous analysis of the *only* shred of evidence for the accepted dogma that they were built by the Pharaoh Cheops and his successors. The chapters exposing a shameless forgery in the Great Pyramid have been the subject of countless magazine articles and media discussions; and the repercussions of our revelations still reverberate.

This book places in context the biblical tales of the pre-Diluvial Enoch and the post-Diluvial prophet Elijah who had been taken up to be with the *Elohim*, and of their Sumerian predecessors Adapa, Etana, and Enmeduranki; by doing that, it converts Man's search for immortality to a realizable, attainable dream.

Zecharia Sitchin
New York, October 2006

1

In Search of Paradise

There was a time—our ancient scriptures tell us—when Immortality was within the grasp of Mankind.

A golden age it was, when Man lived with his Creator in the Garden of Eden—Man tending the wonderful orchard, God taking strolls in the afternoon breeze. "And the Lord God caused to grow from the ground every tree that is pleasant to the sight and good for eating; and the Tree of Life was in the orchard, and the Tree of Knowing good and evil. And a river went out of Eden to water the garden, and from there it was parted and became four principal streams: the name of the first is Pishon . . . and of the second Gihon . . . and of the third Tigris . . . and the fourth river is the Euphrates."

Of the fruit of every tree were Adam and Eve permitted to eat—except of the fruit of the Tree of Knowing. But once they did (tempted by the Serpent)—the Lord God grew concerned over the matter of Immortality:

> Then did the Lord Yahweh say:
> "Behold, the Adam has become as one of us
> to know good and evil;
>   And now might he not put forth his hand
> and partake also of the Tree of Life,
> and eat, and live forever?"

> And the Lord Yahweh expelled the Adam
> from the Garden of Eden . . .

And He placed at the east of the Garden of Eden
the Cherubim, and the Flaming Sword which revolveth,
to guard the way to the Tree of Life.

So was Man cast out of the very place where eternal life
was within his grasp. But though barred from it, he has never
ceased to remember it, to yearn for it, and to try to reach it.

Ever since that expulsion from Paradise, heroes have gone
to the ends of Earth in search of Immortality; a selected few
were given a glimpse of it; and simple folk claimed to have
chanced upon it. Throughout the ages, the Search for Paradise was the realm of the individual; but earlier in this millenium, it was launched as the national enterprise of mighty
kingdoms.

The New World was discovered—so have we been led
to believe—when explorers went seeking a new, maritime
route to India and her wealth. True—but not the whole truth;
for what Ferdinand and Isabel, king and queen of Spain, had
desired most to find was the Fountain of Eternal Youth: a
magical fountain whose waters rejuvenate the old and keep
one young forever, for it springs from a well in Paradise.

No sooner had Columbus and his men set foot in what they
all thought were the islands off India (the "West Indies"),
than they combined the exploration of the new lands with a
search for the legendary Fountain whose waters "made old
men young again." Captured "Indians" were questioned,
even tortured, by the Spaniards, so that they would reveal
the secret location of the Fountain.

One who excelled in such investigations was Ponce
de Leon, a professional soldier and adventurer, who rose
through the ranks to become governor of the part of the island of Hispaniola now called Haiti, and of Puerto Rico. In
1511, he witnessed the interrogation of some captured Indians. Describing their island, they spoke of its pearls and
other riches. They also extolled the marvelous virtues of
its waters. A spring there is, they said, of which an islander
"grievously oppressed with old age" had drunk. As a re-

sult, he "brought home manly strength and has practiced all manly performances, having taken a wife again and begotten children."

Listening with mounting excitement, Ponce de Leon—himself an aging man—was convinced that the Indians were describing the miraculous Fountain of the rejuvenating waters. Their postscript, that the old man who drank of the waters regained his manly strength, could resume practicing "all manly performances," and even took again a young wife who bore him children—was the most conclusive aspect of their tale. For in the court of Spain, as throughout Europe, there hung numerous paintings by the greatest painters, and whenever they depicted love scenes or sexual allegories, they included in the scene a fountain. Perhaps the most famous of such paintings, Titian's *Love Sacred and Love Profane,* was created at about the time the Spaniards were on their quest in the Indies. As everyone well knew, the Fountain in the paintings hinted at the ultimate lovemaking; the Fountain whose waters make possible "all manly performances" through Eternal Youth.

Ponce de Leon's report to King Ferdinand is reflected in the records kept by the official court historian, Peter Martyr de Angleria. As stated in his *Decade de Orbe Novo* [Decades of the New World], the Indians who had come from the islands of Lucayos or the Bahamas, had revealed that "there is an island . . . in which there is a perennial spring of running water of such marvelous virtue, that the waters thereof being drunk, perhaps with some diet, make old men young again." Many researches, such as *Ponce de Leon's Fountain of Youth: History of a Geographical Myth* by Leonardo Olschki, have established that "the Fountain of Youth was the most popular and characteristic expression of the emotions and expectations which agitated the conquerors of the New World." Undoubtedly, Ferdinand the king of Spain was one of those so agitated, so expectant for the definitive news.

So, when word came from Ponce de Leon, Ferdinand lost little time. He at once granted Ponce de Leon a Patent of Discovery (dated February 23, 1512), authorizing an expedition from the island of Hispaniola northward. The admiralty was ordered to assist Ponce de Leon and make available to him the best ships and seamen, so that he might discover without delay the island of "Beininy" (Bimini). The king made one condition explicit: "that after having reached the island and learned what is in it, you shall send me a report of it."

In March 1513, Ponce de Leon set out northward, to look for the island of Bimini. The public excuse for the expedition was a search for "gold and other metals"; the true aim was to find the Fountain of Eternal Youth. This the seamen soon learnt as they came upon not one island but hundreds of islands in the Bahamas. Anchoring at island after island, the landing parties were instructed to search not for gold but for some unusual fountain. The waters of each stream were tasted and drunk—but with no evident effects. On Easter Sunday— *Pasca de Flores* by its Spanish name—a long coastline was sighted. Ponce de Leon called the "island" Florida. Sailing along the coast and landing again and again, he and his men searched the jungled forests and drank the waters of endless springs. But none seemed to work the expected miracle.

The mission's failure appears to have hardly dampened the conviction that the Fountain was undoubtedly there: it only had to be discovered. More Indians were questioned. Some seemed unusually young for the old ages claimed by them. Others repeated legends that confirmed the existence of the Fountain. One such legend (as recounted in *Creation Myths of Primitive America* by J. Curtin) relates that when Olelbis, "He Who Sits Above," was about to create Mankind, he sent two emissaries to Earth to construct a ladder which would connect Earth and Heaven. Halfway up the ladder, they were to set up a resting place, with a pool of pure drinking waters. At the summit, they were to create two springs: one for drinking and the other for bathing. When a man or woman grows old, said Olelbis, let him or her climb up to this summit, and drink and bathe; whereupon his youth shall be restored.

The conviction that the Fountain existed somewhere on the islands was so strong that in 1514—the year after Ponce de Leon's unfruitful mission—Peter Martyr (in his Second Decade) informed Pope Leo X as follows:

> At a distance of 325 leagues from Hispaniola, they tell, there is an island called Boyuca, alias Ananeo, which—according to those who explored its interior—has such an extraordinary fountain that drinking of its waters rejuvenates the old.

> And let Your Holiness not think this to be said lightly or rashly; for they have spread word of this as the truth throughout the court, so formally that the whole people, not few of whom are from among those whom wisdom or fortune distinguished from the common people, hold it to be true.

Ponce de Leon, undaunted, concluded after some additional research that what he had to look for was a spring in conjunction with a river, the two possibly connected by a hidden underground tunnel. If the Fountain was on an island, was its source a river in Florida?

In 1521, the Spanish Crown sent Ponce de Leon on a renewed search, this time focusing on Florida. There can be no doubt regarding the true purpose of his mission: writing only a few decades later, the Spanish historian Antonio de Herrera y Tordesillas stated thus in his *Historia General de las Indias:* "He (Ponce de Leon) went seeking that Sacred Fountain, so renowned among the Indians, as well as the river whose waters rejuvenated the aged." He was intent on finding the spring of Bimini and the river in Florida, of which the Indians of Cuba and Hispaniola "affirmed that old persons bathing themselves in them became young again."

Instead of Eternal Youth, Ponce de Leon found death by an Indian arrow. And although the individual search for a potion or lotion that can postpone the Last Day may never end, the organized search, under a royal decree, did come to an end.

Was the search futile to begin with? Were Ferdinand and Isabel and Ponce de Leon, and the men who sailed and died in search of the Fountain, all fools childishly believing in some primitive fairy tales?

Not the way they saw it. The Holy Scriptures, pagan beliefs, and the documented tales of great travelers, all combined to affirm that there was indeed a place whose waters (or fruits' nectars) could bestow Immortality by keeping one forever young.

There were still current olden tales—left from the times when the Celts were in the peninsula—of a secret place, a secret fountain, a secret fruit or herb whose finder shall be redeemed of death. There was the goddess Idunn, who lived by a sacred brook and who kept magical apples in her coffer. When the gods grew old, they would come to her to eat of the apples, whereupon they turned young again. Indeed, "Idunn" meant "Again Young"; and the apples that she guarded were called the "Elixir of the Gods."

Was this an echo of the legend of Herakles (Hercules) and his twelve labors? A priestess of the god Apollo, predicting his travails in an oracle, had also assured him: "When this shall be done, thou shalt become one of the Immortals." To achieve this, the last but one labor was to seize and bring back from the Hesperides the divine golden apples. The Hesperides—"Daughters of the Evening Land"—resided at the Ends of Earth.

Have not the Greeks, and then the Romans, left behind them tales of mortals immortalized? The god Apollo anointed the body of Sarpedon, so that he lived the life of several generations of men. The goddess Aphrodite granted to Phaon a magic potion; anointing himself with it, he turned into a beautiful youth "who wakened love in the hearts of all the women of Lesbos." And the child Demophon, anointed with ambrosia by the goddess Demeter, would surely have become immortal were not his mother—ignorant of Demeter's identity—to snatch him away from the goddess.

There was the tale of Tantalus, who had become immortal by eating at the gods' table and stealing their nectar and ambrosia. But having killed his son to serve his flesh as food for the gods, he was punished by being banished to a land of luscious fruits and waters—eternally out of his reach. (The god Hermes restored the butchered son to life.) On the other hand, Odysseus, offered Immortality by the nymph Calypso if only he would stay with her forever, forsook Immortality for a chance to return to his home and wife.

And was not there the tale of Glaukos, a mortal, an ordinary fisherman, who became a sea-god? One day he observed that a fish that he had caught, coming in touch with a herb, came back to life and leaped back into the water. Taking the herb into his mouth, Glaukos jumped into the water at the exact same spot; whereupon the sea-gods Okeanos and Tethys admitted him to their circle and transformed him into a deity.

The year 1492, in which Columbus set sail from Spain, was also the year in which the Muslim occupation of the Iberian Peninsula ended with the surrender of the Moors at Granada. Throughout the nearly eight centuries of Muslim and Christian contention over the peninsula, the interaction of the two cultures was immense; and the tale in the Koran (the Muslim holy book) of the Fish and the Fountain of Life was known to Moor and Catholic alike. The fact that the tale was almost identical to the Greek legend of Glaukos the fisherman, was taken as confirmation of its authenticity. It was one of the reasons for seeking the legendary Fountain in India—the land which Columbus had set out to reach, and which he thought he had reached.

The segment in the Koran which contains the tale is the eighteenth *Sura*. It relates the exploration of various mysteries by Moses, the biblical hero of the Israelite Exodus from Egypt. While Moses was being groomed for his new calling as a Messenger of God, he was to be instructed in such knowledge as he still lacked by a mysterious "Servant of God." Accompanied by only one attendant, Moses was to go find this enigmatic teacher with the aid of a single clue: he was to take with him a dried fish; the place where the fish would jump and disappear would be the place where he would meet the teacher.

After much searching in vain, the attendant of Moses suggested that they stop and give up the search. But Moses persisted, saying that he would not give up until they reached "the junction of the two streams." Unnoticed by them, it was there that the miracle happened:

> But when they reached the Junction,
> they forgot about their fish,
> which took its course through the stream,
> as in a tunnel.

After journeying further, Moses said to his attendant: "Bring us our early meal." But the attendant replied that the fish was gone:

> "When we betook ourselves to the rock,
> sawest thou what had happened?
> I did indeed forget about the fish—
> Satan made me forget to tell you about it:
> It took its course through the stream,
> in a marvelous way.
> And Moses said:
> "That was what we were seeking after."

The tale in the Koran (Fig. 1) of the dried fish that came to life and swam back to the sea through a tunnel, went beyond the parallel Greek tale by relating itself not to a simple fisherman, but to the venerated Moses. Also, it presented the incident not as a chance discovery, but as an occurrence premediated by the Lord, who knew of the location of the

60. Behold, Moses said
To his attendant, " I will not
Give up until I reach
The junction of the two
Seas or (until) I spend
Years and years in travel."

٦٠- وَإِذْ قَالَ مُوسَى لِفَتَـٰهُ لَآ
أَبْرَحُ حَتَّىٰ أَبْلُغَ مَجْمَعَ الْبَحْرَيْنِ أَوْ
أَمْضِيَ حُقُبًا ۝

61. But when they reached
The Junction, they forgot
(About) their Fish, which took
Its course through the sea
(Straight) as in a tunnel.

٦١- فَلَمَّا بَلَغَا مَجْمَعَ بَيْنِهِمَا نَسِيَا حُوتَهُمَا
فَاتَّخَذَ سَبِيلَهُ فِى الْبَحْرِ سَرَبًا ۝

62. When they had passed on
(Some distance), Moses said
To his attendant : " Bring us
Our early meal; truly
We have suffered much fatigue
At this (stage of) our journey."

٦٢- فَلَمَّا جَاوَزَا قَالَ
لِفَتَـٰهُ ءَاتِنَا غَدَآءَنَا لَقَدْ
لَقِينَا مِن سَفَرِنَا هَـٰذَا نَصَبًا ۝

63. He replied : " Sawest thou
(What happened) when we
Betook ourselves to the rock ?
I did indeed forget
(About) the Fish : none but
Satan made me forget
To tell (you) about it :
It took its course through
The sea in a marvellous way ! "

٦٣- قَالَ أَرَءَيْتَ إِذْ أَوَيْنَآ
إِلَى الصَّخْرَةِ فَإِنِّى نَسِيتُ الْحُوتَ
وَمَآ أَنسَىٰنِيهُ إِلَّا الشَّيْطَـٰنُ أَنْ أَذْكُرَهُ
وَاتَّخَذَ سَبِيلَهُ فِى الْبَحْرِ عَجَبًا ۝

64. Moses said : " That was what
We were seeking after: "
So they went back
On their footsteps, following
(The path they had come).

٦٤- قَالَ ذَٰلِكَ مَا كُنَّا نَبْغِ
فَارْتَدَّا عَلَىٰ ءَاثَارِهِمَا قَصَصًا ۝

Fig. 1

Waters of Life—waters that could be recognized through the
medium of the resurrection of a dead fish.

As devout Christians, the king and queen of Spain must
have accepted literally the vision described in the Book of
Revelation, "of a pure river of Water of Life, clear as crystal, proceeding out of the throne of God. . . . In the midst of
the street of it, and on either side of the river, was there the

Tree of Life, with twelve manner of fruit." They must have believed in the Book's promises: "I will give unto him that is athirst of the Fountain of the Water of Life"—"I will give to eat of the Tree of Life which is the midst of the Paradise of God." And could they not be aware of the words of the biblical Psalmist:

> Thou givest them to drink
> of thy Stream of Eternities;
> For with thee is the Fountain of Life.

There could thus be no doubt, as attested by the holiest Scriptures, that the Fountain of Life, or the Stream of Eternity, did exist; the only problem was—where, and how to find it.

The eighteenth *Sura* of the Koran seemed to offer some important clues. It goes on to relate the three paradoxes of life that Moses was shown once he located the Servant of God. Then the same section of the Koran continues to describe three other episodes: first, about a visit to a land where the Sun sets; then to a land where the Sun rises—that is, in the east; and finally to a land beyond the second land, where the mythical people of Gog and Magog (the biblical contenders at the End of Days) were causing untold mischief on Earth. To put an end to this trouble, the hero of the tale, here named Du-al'karnain ("Possessor of the Two Horns"), filled up the pass between two steep mountains with blocks of iron and poured over them molten lead, creating such an awesome barrier that even the mighty Gog and Magog were powerless to scale it. Separated, the two could cause no more hardship on Earth.

The word Karnain, in Arabic as in Hebrew, means both Double Horns and Double Rays. The three additional episodes, following immediately after the tale of the Mysteries of Moses, thus appear to retain as their hero Moses, who could well have been nicknamed Du-al'karnain because his face "was with rays"—radiated—after he had come down from Mount Sinai, where he had met the Lord face to face. Yet popular medieval beliefs attributed the epithet and the journeys to the three lands to Alexander the Great, the Macedonian king who in the fourth century B.C. conquered most of the ancient world, reaching as far as India.

This popular belief, interchanging Moses and Alexander, stemmed from traditions concerning the conquests and adventures of Alexander the Great. These included not only the feat in the land of Gog and Magog, but also an identical episode of a dry, dead fish that came back to life when Alexander and his cook had found the Fountain of Life!

The reports concerning Alexander's adventures that were current in Europe and the Near East in medieval times were based upon the supposed writings of the Greek historian Callisthenes of Olynthus. He was appointed by Alexander to record the exploits, triumphs and adventures of his Asiatic expedition; but he died in prison, having offended Alexander, and his writings have mysteriously perished. Centuries later, however, there began to circulate in Europe a Latin text purporting to be a translation of the lost original writings of Callisthenes. Scholars speak of this text as "pseudo-Callisthenes."

For many centuries it was believed that the many translations of the Exploits of Alexander that were current in Europe and the Middle East, all stemmed from this Latin pseudo-Callisthenes. But it was later discovered that other, parallel versions existed in many languages—including Hebrew, Arabic, Persian, Syriac, Armenian and Ethiopic—as well as at least three versions in Greek. The various versions, some tracing their origins to Alexandria of the second century B.C., differ here and there; but by and large, their overwhelming similarities do indicate a common source—perhaps the writings of Callisthenes after all, or—as is sometimes claimed—copies of Alexander's letters to his mother Olympias and to his teacher Aristotle.

The miraculous adventures with which we are concerned began after Alexander completed the conquest of Egypt.

From the texts it is neither clear in which direction Alexander set his course, nor is it certain that the episodes are arranged in an accurate chronological or geographical order. One of the very first episodes, however, may explain the popular confusion between Alexander and Moses: apparently Alexander attempted to leave Egypt as Moses did, by parting the waters and getting his followers to cross the sea on foot.

Reaching the sea, Alexander decided to part it by building in its midst a wall of molten lead, and his masons "continued to pour lead and molten matter into the water until the structure rose above its surface. Then he built upon it a tower and a pillar, upon which he carved his own figure, having two horns upon his head." And he wrote upon the monument: "Whosoever hath come into this place and would sail over the sea, let him know that I have shut it up."

Having thus shut out the waters, Alexander and his men began to cross the sea. As a precaution, however, they sent ahead some prisoners. But as they reached the tower in the midst of the waters, "the waves of the sea leapt up upon them (the prisoners) and the sea swallowed them up and they all perished. . . . When the Two-Horned One saw this, he was afraid of the sea with a mighty fear," and gave up the attempt to emulate Moses.

Eager, however, to discover the "darkness" on the other side of the sea, Alexander made several detours, during which he purportedly visited the sources of the river Euphrates and of the river Tigris, studying there "the secrets of the heavens and the stars and the planets."

Leaving his troops behind, Alexander returned toward the Land of Darkness, reaching a mountain named *Mushas* at the edge of the desert. After several days of marching, he saw "a straight path which had no wall, and it had no high and no low place in it." He left his few trusted companions and proceeded alone. After a journey of twelve days and twelve nights, "he perceived the radiance of an angel"; but as he drew nearer, the angel was "a flaming fire." Alexander realized that he had reached "the mountain from which the whole world is surrounded."

The angel was no less puzzled than Alexander. "Who art thou, and for what reason art thou here, O mortal?" the angel asked, and wondered how Alexander had managed "to penetrate into this darkness, which no other man hath been able to do." To which Alexander replied that God himself had guided him and gave him strength to "have arrived in this place, which is Paradise."

To convince the reader that Paradise, rather than Hell, was reachable through underground passages, the ancient author then introduced a long discourse between the angel and Alexander on matters of God and Man. The angel then urged Alexander to return to his friends; but Alexander persisted in seeking answers to the mysteries of Heaven and Earth, God and Men. In the end Alexander said that he would leave only if he were granted something that no man had ever obtained before. Complying, "the angel said unto him: 'I will tell thee something whereby thou mayest live and not die.' The Two-Horned said: 'Say on.' And the angel said unto him:

> In the land of Arabia, God hath set the blackness of solid darkness, wherein is hidden a treasury of this knowledge. There too is the fountain of water which is called "The Water of Life"; and whosoever drinketh therefrom, if it be but a single drop, shall never die.

The angel attributed other magical powers to these Waters of Life, such as "the power of flying through the heavens, even as the angels fly." Needing no further prompting, Alexander anxiously inquired: "In which quarter of the earth is this fountain of water situated?" "Ask those men who are heirs to the knowledge thereof," was the angel's enigmatic answer. Then the angel gave Alexander a cluster of grapes whereby to feed his troops.

Returning to his companions, Alexander told his colleagues of his adventure and gave them each a grape. But "as

he plucked off from the cluster, another grew in its place."
And so did one cluster feed all the soldiers and their beasts.

Alexander then began to make inquiries of all the learned
men he could find. He asked the sages: "Have you ever read
in your books that God hath a place of darkness of which
knowledge is hidden, and that the Fountain which is called
the 'Fountain of Life' is situated therein?" The Greek versions have him search the Ends of Earth for the right savant;
the Ethiopic versions suggest that the sage was right there,
among his troops. His name was Matun, and he knew the
ancient writings. The place, he said, "lieth nigh unto the sun
when it rises on the right side."

Scarcely more informed by such riddles, Alexander put
himself in the hands of his guide. They again went into a
Place of Darkness. After journeying for a long time, Alexander got tired and sent Matun ahead by himself, to find the
right path. To help him see in the darkness, Alexander gave
him a stone which was given him earlier under miraculous
circumstances by an ancient king who was living among the
gods—a stone which was brought out of Paradise by Adam
when he left it, and which was heavier than any other substance on Earth.

Matun, though careful to follow the path, eventually lost
his way. He then produced the magical stone and put it down;
when it touched the ground, it emitted light. By the light,
Matun saw a well. He was not yet aware that he had chanced
upon the Fountain of Life. The Ethiopic version describes
what ensued:

> Now, he had with him a dried fish, and being exceed-
> ingly hungry he went down with it to the water, that
> he might wash it therein and make it ready for cook-
> ing. . . . But behold, as soon as the fish touched the
> water, it swam away.

"When Matun saw this, he stripped his clothes and went
down into the water after the fish, and found it to be alive in

the water." Realizing that this was the "Well of the Water of Life," Matun washed himself in the waters and drank thereof. When he had come up from the well, he was no longer hungry nor did he have any worldly care, for he had become *El-Khidr*—"the Evergreen"—the one who was Forever Young.

Returning to the encampment, he said nothing of his discovery to Alexander (whom the Ethiopic version calls "He of the Two Horns"). Then Alexander himself resumed the search, groping for the right way in the darkness. Suddenly, he saw the stone (left behind by Matun) "shining in the darkness; (and) now it had two eyes, which sent forth rays of light." Realizing that he had found the right path, Alexander rushed ahead, but was stopped by a voice which admonished him for his ever-increasing ambitions, and prophesied that instead of eternal life he would soon bite the dust. Terrified, Alexander returned to his companions and his troops, giving up the search.

According to some versions, it was a bird with human features who spoke to Alexander and made him turn back when he reached a place "inlaid with sapphires and emeralds and jacinths." In Alexander's purported letter to his mother, there were two bird-men who blocked his way.

In the Greek version of pseudo-Callisthenes, it was Andreas, the cook of Alexander, who took the dried fish to wash it in a fountain "whose waters flashed with lightnings." As the fish touched the water, it became alive and slipped out of the cook's hands. Realizing what he had found, the cook drank of the waters and took some in a silver bowl—but told no one of his discovery. When Alexander (in this version, he was accompanied by 360 men) continued the search, they reached a place that shined though there were neither Sun nor Moon nor stars to be seen. The way was blocked by two birds with human features.

"Go back!" one of them ordered Alexander, "for the land on which you stand belongs to God alone. Go back, O wretched one, for in the Land of the Blessed you cannot set foot!" Shuddering with fear, Alexander and his men turned back; but before they left the place, they took as souvenirs some of its soil and stones. After several days' marching, they came out of the Land of Everlasting Night; and when they reached light, they saw that the "soil and stones" they picked up were in fact pearls, precious stones and nuggets of gold.

Only then did the cook tell Alexander of the fish that came to life, but still kept it a secret that he himself had drunk of the waters and that he had kept some of it. Alexander was furious and hit him, and banished him from the camp. But the cook wished not to leave alone, for he had fallen in love with a daughter of Alexander. So he revealed his secret to her, and gave her to drink of the waters. When Alexander found that out, he banished her too: "You have become a godly being, having become immortal," he told her; therefore, he said, you cannot live among men—go live in the Land of the Blessed. And as for the cook—him Alexander threw into the sea, with a stone around his neck. But instead of drowning, the cook became the sea-demon Andrentic.

"And thus," we are told, "ends the tale of the Cook and the Maiden."

To the learned advisers of Europe's medieval kings and queens, the various versions only served to confirm both the antiquity and the authenticity of the legend of Alexander and the Fountain of Life. But where, O where were these magical waters located?

Were they indeed across the border of Egypt, in the Sinai peninsula—the arena of the activities of Moses? Were they close to the area where the Euphrates and Tigris rivers begin to flow, somewhere north of Syria? Did Alexander go to the Ends of Earth—India—to find the Fountain, or did he embark on those additional conquests after he was turned back from it?

As the medieval scholars strove to unravel the puzzle, new works on the subject from Christian sources began to shape a consensus in favor of India. A Latin composition named *Alexander Magni Inter Ad Paradisum,* a Syriac Homilie of Alexander by Bishop Jakob of Sarug, the *Recension of Josippon* in Armenian—complete with the tale of the tunnel, the man-like birds, the magical stone—placed the Land of Darkness or the Mountains of Darkness at the Ends of Earth. There, some of these writings said, Alexander took a boat ride on the Ganges River, which was none other than the Pishon River of Paradise. There, in India (or on an island offshore), did Alexander reach the Gates of Paradise.

As these conclusions were taking shape in Europe of the Middle Ages, new light was shed on the subject from a wholly unexpected source. In the year 1145, the German bishop Otto of Freising reported in his *Chronicon* a most astonishing epistle. The Pope, he reported, had received a letter from a Christian ruler of India, whose existence had been totally unknown until then. And that king had affirmed in his letter that the River of Paradise was indeed located in his realm.

Bishop Otto of Freising named as the intermediary, through whom the Pope had received the epistle, Bishop Hugh of Gebal, a town on the Mediterranean coast of Syria. The ruler, it was reported, was named John the Elder or, being a priest, Prester John. He was reputedly a lineal descendant of the Magi who had visited Christ the child. He defeated the Muslim kings of Persia, and formed a thriving Christian kingdom in the lands of the Ends of Earth.

Nowadays, some scholars consider the whole affair to have been a forgery for propaganda purposes. Others believe that the reports which reached the Pope were distortions of events that were really happening. The Christian world at the time, having launched the Crusades against Muslim rule over the Near East (including the Holy Land) fifty years earlier, met with a crushing defeat at Edessa in 1144. But at the Ends of Earth Mongol rulers began to storm the gates of the Muslim empire, and in 1141 defeated the Sultan Sanjar. When the news reached the Mediterranean coastal cities, it was forwarded to the Pope in the garb of a Christian king, rising to defeat the Muslims from the other side.

If the search for the Fountain of Youth was not among the reasons for the First Crusade (1095), it apparently was among the reasons for the following ones. For no sooner had Bishop Otto reported the existence of Prester John and of the River of Paradise in his realm, than the Pope issued a formal call for the resumption of the Crusades. Two years later, in 1147, Emperor Conrad of Germany, accompanied by other rulers and many nobles, launched the Second Crusade.

As the fortunes of the Crusaders rose and fell, Europe was swept anew by word from Prester John and his promises of aid. According to chroniclers of those days, Prester John sent in 1165 a letter to the Byzantine emperor, to the Holy Roman emperor, and to lesser kings, in which he declared his definite intention to come to the Holy Land with his armies. Again his realm was described in glowing terms, as befits the place where the River of Paradise—indeed, the Gates of Paradise—were situated.

The promised help never came. The way from Europe to India was not breached open. By the end of the thirteenth century, the Crusades were over, ending in final defeat at the hands of the Muslims.

But even as the Crusaders were advancing and retreating, the fervent belief in the existence of the Waters of Paradise in India kept growing and spreading.

Before the twelfth century was over, a new and popular version of the exploits of Alexander the Great made its way into the encampments and town squares. Called the *Romance of Alexander,* it was (as is now known) the work of two Frenchmen who based this poetic and glowing composition on the Latin version of pseudo-Callisthenes and other "biographies" of the Macedonian hero then available. The knights, the warriors, the townspeople in the drinking halls, cared not who the authors were; for—in language they could understand—it vividly drew for them visions of Alexander's adventures in strange lands.

Among these was the tale of the three wondrous fountains. One rejuvenated the old; the second granted Immortality; the third resurrected the dead. The three fountains, the *Romance* explained, were located in different lands, issuing as they were from the Tigris and Euphrates rivers in western Asia, the Nile in Africa, and the Ganges River in India. These were the four Rivers of Paradise; and though they flowed in different lands, they all arose from a single source: from the Garden of Eden—just as the Bible had stated all along.

It was the Fountain of Rejuvenation, the *Romance* related, that Alexander and his men had found. It recounted as fact that fifty-six aged companions of Alexander "recovered the complexion of thirty [years] old after drinking from the Fountain of Youth." As translations of the *Romance* carried the tale far and wide, the versions became increasingly specific on this point: not only the appearance, but also the manhood and virility of the aged soldiers were restored to youthfulness.

But how does one get to this Fountain, if the route to India is blocked by the heathen Muslims?

On and off, the Popes attempted to communicate with the enigmatic Prester John, "the illustrious and magnificent king of the Indies and beloved son of Christ." In 1245, Pope Innocent IV dispatched the Friar Giovanni da Pian del Carpini, via southern Russia, to the Mongol ruler or Khan, believing the Mongols to be Nestorians (an offshoot of the Eastern Orthodox Church) and the Khan to be Prester John. In 1254, the Armenian ruler-priest Haithon traveled in disguise through eastern Turkey to the camp of the Mongol chieftain in southern Russia. The record of his adventurous travels mentioned that his way took him via the narrow pass on the shores of the Caspian Sea called *The Iron Gates;* and the speculation that his route resembled that of Alexander the Great (who had poured molten iron to close a mountain pass) only served to suggest that the Ends of Earth, the Gates of Paradise, could indeed be so reached.

These and other papal and royal emissaries were soon

joined by private adventurers, such as the brothers Nicolo and Maffeo Polo and the former's son Marco Polo (1260–1295), and the German knight William of Boldensele (1336)—all searching for the kingdom of Prester John.

While their travelogues kept up the interest of Church and Courts, it was once again the fate of a popular literary work to rekindle mass interest. Its author introduced himself as "I, John Maundeville, Knight," born in the town of St. Albans in England who "passed the sea in the year of our Lord Jesus 1322." Writing at the end of his travels thirty-four years later, Sir John explained that he had therein "set down the way to the Holy Land, and to Hierusalem: as also to the lands of the Great Caan, and of Prester John: to Inde, and divers other countries: together with many and strange marvels therein."

In the twenty-seventh chapter, captioned "Of the Royal Estate of Prester John," the book *(The Voyages and Travels of Sir John Maundeville, Knight)* states:

> This emperor, Prester John, possesses very extensive territory, and has many noble cities and good towns in his realm, and many great and large isles. For all the country of India is divided into isles, by the great floods that come from Paradise. . . .
> And this land is full good and rich. . . . In the land of Prester John are many divers things and many precious stones, so great and so large, that men make thereof plates, dishes, cups etc. . . .

Sir John went on to describe the River of Paradise:

> In his country is the sea called the Gravelly Sea. . . . Three days from that sea are great mountains, out of which runs a great river which comes from Paradise, and it is full of precious stones, without a drop of water, and it runs through the desert, on one side, so that it makes the Gravelly Sea where it ends.

Beyond the River of Paradise, there was "a great isle, long and broad, called Milsterak," that was a paradise on Earth. It had "the fairest garden that might be imagined; and therein were trees bearing all manner of fruits, all kinds of herbs of virtue and of good smell." This paradise, Sir John states, had marvelous pavilions and chambers, the purpose of which was diverse sexual enjoyment, all the work of a rich and devilish man.

Having fired the imagination (and greed) of his readers with the tales of precious stones and other riches, Sir John now played on the men's sexual desires. The place, he wrote, was filled with "the fairest damsels that might be found under the age of fifteen years, and the fairest young striplings that men might get of that same age, and they were all clothed richly in clothes of gold; and he said that they were angels." And the devilish man—

> Had also caused to be made three fair and noble wells, all surrounded with stone of jasper and crystal, diapered with gold, and set with precious stones and great Orient pearls. And he had made a conduit under the earth, so that the three wells, at his will, should run one with milk, another with wine, and another with honey. And that place he called Paradise.

To that place, the crafty man lured "good knights, hardy and noble," and after entertaining them he persuaded them to go and kill his enemies; telling them that they should not fear being slain, for should they die, they would be resurrected and rejuvenated:

> After their death they should come to his Paradise, and they should be of the age of the damsels, and they should play with them. And after that he would put them in a fairer Paradise, where they should see the God of Nature visibly, in his majesty and bliss.

But that, said John Maundeville, was not the real Paradise of biblical renown. That one, he said in Chapter XXX, lay beyond the isles and lands which Alexander the Great had journeyed through. The route to it led farther east, toward two isles rich in gold and silver mines "where the Red Sea separates from the Ocean Sea":

> And beyond that land and isles, and deserts of Prester John's lordship, in going straight toward the east, men find nothing but mountains and great rocks; and there is the dark region, where no man can see, neither by day nor night. . . . And that desert, and that place of darkness, lasts from this coast unto Terrestrial Paradise, where Adam, our first father, and Eve were put.

It was from there that the waters of Paradise flowed:

> And in the highest place of Paradise, exactly in the middle, is a well that casts out the four streams, which run by diverse lands, of which the first is called Pison, or Ganges, that runs through India, or Emlak, in which river are many precious stones, and much lignum aloes, and much sand of gold.
> And the other river is called Nile, or Gyson, which goes through Ethiopia, and after through Egypt.
> And the other is called Tigris, which runs by Assyria, and by Armenia the Great.
> And the other is called Euphrates, which runs through Media, Armenia and Persia.

Confessing that he himself had not reached this biblical Garden of Eden, John Maundeville explained: "No mortal man may approach to that place without special grace of God; so that of that place I can tell you no more."

In spite of this admission, the many versions in many languages that flowed from the English original maintained that the knight had stated "I, John de Maundeville, saw that

Fountain and drank three times of that water with my companion, and since I drank I feel well." The fact that in the English version, Maundeville complained that he was sick with rheumatic gout and near the end of his days, mattered not to the many who were thrilled by the marvelous tales. Nor did it matter then, that scholars nowadays believe that "Sir John Maundeville, Knight" may in fact have been a French doctor who had never traveled, but very skillfully put together a travelogue from the writings of others who did take the risk and trouble of journeying far and away.

Writing about the visions that had motivated the exploration that led to the discovery of America, Angel Rosenblat *(La Primera Vision de America y Otros Estudios)* summed up the evidence thus: "Along with the belief in the earthly Paradise was associated another desire of a messianic (or Faustic) nature; to find the Fountain of Eternal Youth. All the Middle Ages had dreamed of it. In the new images of the Lost Paradise, the Tree of Life was converted into the Fountain of Life, and then into a River or Spring of Youth." The motivation was the conviction that "the Fountain of Life came from India . . . a Fountain that cured all ills and assured immortality. The fantastic John Maundeville had actually encountered it on his trip to India . . . in the Christian Kingdom of Prester John." To reach India and the waters that flow from Paradise became "a symbol of the eternal human desire for pleasure, youth and happiness."

With the land routes blocked by enemies, the Christian kingdoms of Europe sought a sea route to India. Under Henry the Navigator, the kingdom of Portugal emerged in the middle of the fifteenth century as the leading power in the race to reach the Orient by sailing around Africa. In 1445, the Portuguese navigator Dinas Dias reached the mouth of the Senegal River, and mindful of the voyage's purpose reported that "men say it comes from the Nile, being one of the most glorious rivers of Earth, flowing from the Garden of Eden and the earthly Paradise." Others followed, pushing to and around the Cape at the tip of the African continent. In 1499,

Vasco da Gama and his fleet circumnavigated Africa and reached the cherished target: India.

Yet the Portuguese, who had launched the Age of Discovery, failed to win the race. Diligently studying the ancient maps and all the writings of those who had ventured east, an Italian-born seaman named Cristóbal Colón concluded that by sailing *west,* he could reach India by a sea route much shorter than the Eastern Route sought by the Portuguese. Seeking a sponsor, he arrived at the court of Ferdinand and Isabel. He had with him (and took on his first voyage) an annotated copy of the Latin version of Marco Polo's book. He could also point to the writings of John Maundeville, who explained a century and a half before Columbus (Colón) that by going to the farthest east, one arrives at the west "on account of the roundness of the earth . . . for our Lord God made the earth all around."

In January 1492, Ferdinand and Isabel defeated the Muslims and expelled them from the Iberian Peninsula. Was it not a divine sign to Spain, that what the Crusaders could not achieve, Spain would? On August 3 of the same year, Columbus sailed under the Spanish flag to find a western sea route to India. On October 12, he sighted land. Until his death in 1506, he was sure that he had reached the islands which made up a great part of the legendary domain of Prester John.

Two decades later, Ferdinand issued to Ponce de Leon the Patent of Discovery, instructing him to find without delay the rejuvenating waters.

The Spaniards had thought that they were emulating Alexander the Great. Little did they know that they were following footsteps of far greater antiquity.

II

The Immortal Ancestors

The short life of Alexander the Macedonian—he died at age thirty-three in Babylon—was filled with conquest, adventure, exploration; a burning desire to reach the Ends of Earth, to unravel divine mysteries.

It was not an aimless search. Son of Queen Olympias and presumably of her husband King Philip II, he was tutored by the philosopher Aristotle in all manner of ancient wisdom. Then he witnessed quarreling and divorce between his parents, leading to the flight of his mother with the young Alexander. There was reconciliation, then murder; the assassination of Philip led to the crowning of Alexander when twenty years old. His early military expeditions brought him to Delphi, seat of the renowned oracle. There he heard the first of several prophesies predicting for him fame—but a very short life.

Undaunted, Alexander set out—as the Spaniards did nearly 1,800 years later—to find the Waters of Life. To do so, he had to open the way to the East. It was from there that the gods had come: the great Zeus, who swam across the Mediterranean, from the Phoenician city of Tyre to the island of Crete; Aphrodite who also came from across the Mediterranean, via the island of Cyprus; Poseidon, who brought with him the horse from Asia Minor; Athena, who carried to Greece the olive tree from the lands of western Asia. There, too, according to the Greek historians, whose writings Alexander studied, were the Waters which kept one forever young.

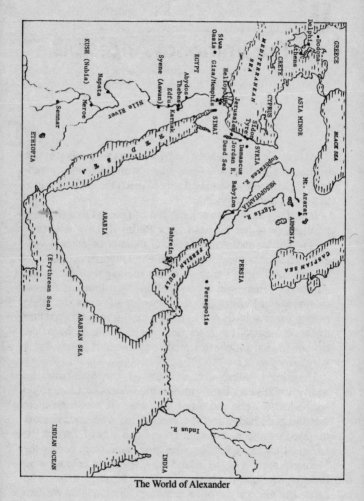

The World of Alexander

Fig. 2

There was the history of Cambyses, son of the Persian king Cyrus, who went by way of Syria, Palestine and the Sinai to attack Egypt. Defeating the Egyptians, he treated them cruelly, and defiled the temple of their god Ammon. Then he took into his heart to go south and attack "the long-lived Ethiopians." Describing the events, Herodotus—writing a century before Alexander—said (*History,* Book III):

> His spies went to Ethiopia, under the pretense of carrying presents to the king, but in reality to take note of all they saw, and especially to observe whether there was really what is called "The Table of the Sun" in Ethiopia. . . .

Telling the Ethiopian king that "eighty years was the longest term of man's life among the Persians," the spies/emissaries questioned him regarding the rumored long life of the Ethiopians. Confirming this,

> The king led them to a fountain, wherein when they had washed, they found their flesh all glossy and sleek, as if they had bathed in oil. And a scent came from the spring like that of violets.

Returning to Cambyses, the spies described the water as "so weak, that nothing would float on it, neither wood nor any lighter substance, but all went to the bottom." And Herodotus noted the following conclusion:

> If the account of this fountain be true, it would be their (the Ethiopians') constant use of the water from it, which makes them so long-lived.

The tale of the Fountain of Youth in Ethiopia, and of the violation by the Persian Cambyses of the temple of Ammon, had direct bearing on the history of Alexander. This concerned the rumors that he was not really the son of Philip, but the offspring of a union between his mother Olympias and the Egyptian god Ammon (Fig. 3). The strained relations between Philip and Olympias only served to confirm the suspicions.

As related in various versions of pseudo-Callisthenes, the court of Philip was visited by an Egyptian Pharaoh whom the Greeks called Nectanebus. He was a master magician, a diviner; and he secretly seduced Olympias. Unbeknown to her at the time, it was in reality the god Ammon who had come to her, taking the guise of Nectanebus. And so it was that when she bore Alexander, she gave birth to a son of a god. It was the very god whose temple the Persian Cambyses had desecrated.

Defeating the Persian armies in Asia Minor, Alexander turned toward Egypt. Expecting heavy resistance by the Persian viceroys who ruled Egypt, he was astonished to see that great land fall into his hands without any resistance: an omen, no doubt. Losing no time, Alexander went to the Great Oasis, seat of the oracle of Ammon. There, the god himself (so legends say) confirmed Alexander's true parentage. Thus affirmed, the Egyptian priests deified him as a Pharaoh; thereby, his desire to escape a mortal's fate became not a privilege, but a right. (Henceforth, Alexander was depicted on his coins as a horned Zeus-Ammon—Fig. 4.)

Alexander then went south to *Karnak,* the center of the worship of Ammon. There was more to the trip than met the eye. A venerated religious center since the third millenium B.C., Karnak was a conglomeration of temples, shrines and monuments to Ammon built by generations of Pharaohs. One of the most impressive and colossal structures was the temple built by Queen Hatshepsut more than a thousand years before Alexander's time. And she too was said to have been a daughter of the god Ammon, conceived by a queen whom the god had visited in disguise!

Whatever actually transpired there, no one really knows. The fact is that instead of leading his armies back east, toward the heartland of the Persian empire, Alexander se-

Fig. 3   Fig. 4

lected a small escort and a few companions for an expedition even farther south. His puzzled companions were led to believe that he was going on a pleasure trip—the pleasures of lovemaking.

The uncharacteristic interlude was as incomprehensible to the historians of those days as to the generals of Alexander. Trying to rationalize, the recorders of Alexander's adventures described the woman he was about to visit as a *femme fatale,* one "whose beauty no living man could praise sufficiently." She was Candace, queen of a land to the south of Egypt (today's Sudan). Reversing the tale of Solomon and the Queen of Sheba, in this instance it was the king who traveled to the queen's land. For, unbeknown to his companions, Alexander was really seeking not love, but the secret of Immortality.

After a pleasant stay, the queen agreed to reveal to Alexander, as a parting gift, the secret of "the wonderful cave where the gods congregate." Following her directions, Alexander found the sacred place:

> He entered with a few soldiers, and saw a starlit haze. And the rooftops were shining, as if lit by stars. The external forms of the gods were physically manifest; a crowd was serving (them) in silence.
>
> At first he (Alexander) was frightened and surprised. But he stayed to see what would happen, for he saw some reclining figures whose eyes were shining like beams of light.

The sight of the "reclining figures," with eyes emitting beams of light, made Alexander stop short. Were they too gods, or deified mortals? He was then startled by a voice: one of the "figures" had spoken up:

> And there was one who said: "Glad greetings, Alexander. Do you know who I am?"
> And he (Alexander) said: "No, my lord."
> The other said: "I am Sesonchusis, the world-conquering king who has joined the ranks of the gods."

Alexander was far from being surprised—as though he had encountered the very person he had searched for. His arrival apparently expected, Alexander was invited in, to "the Creator and Overseer of the entire universe." He "went within, and saw a fire-bright haze; and, seated on a throne, the god whom he had once seen worshipped by men in Rokôtide, the Lord Serapis." (In the Greek version, it was the god Dionysus.)

Alexander saw his chance to bring up the matter of his longevity. "Lord god," he said, "how many years shall I live?"

But there was no answer from the god. Then Sesonchusis sought to console Alexander, for the god's silence spoke for itself. Though I myself have joined the ranks of the gods, Sesonchusis said, "I was not as fortunate as you . . . for although I have conquered the whole world and subjugated so many peoples, nobody remembers my name; but you shall have great renown . . . you will have an immortal name even after death." In this manner, he consoled Alexander. "You shall live upon dying, thus not dying"—immortalized by a lasting reputation.

Disappointed, Alexander left the caves and "continued the journey to be made"—to seek the advice of other sages, to find an escape from a mortal's fate, to emulate others who before him did succeed in joining the immortal gods.

According to one version, among those whom Alexander searched out and met was Enoch, the biblical patriarch from the days before the Deluge, who was the great-grandfather of Noah. It was a place of mountains, "where Paradise, which is the Land of the Living, is situated," the "abode where the saints dwell." Atop a mountain there was a glittering structure, from which there extended skyward a huge stairway, made of 2,500 golden slabs. In a vast hall or cavern Alexander saw "golden figures, each standing in its niche," a golden altar, and two huge "candlesticks" measuring some sixty feet in height.

Upon a couch nearby reclined the form of a man who was draped in a coverlet inlaid with gold and precious stones, and above it, worked in gold, were branches of a vine, having its cluster of grapes formed of jewels.

The man suddenly spoke up, identifying himself as Enoch. "Do not pry into the mysteries of God," the voice warned Alexander. Heeding the warning, Alexander left to rejoin his troops; but not before receiving as a parting gift a bunch of grapes that miraculously were sufficient to feed his whole army.

In yet another version, Alexander encountered not one but two men from the past: Enoch, and the Prophet Elijah—two who according to biblical traditions have never died. It happened when Alexander was traversing an uninhabited desert. Suddenly, his horse was seized by a "spirit" which carried horse and rider aloft, bringing Alexander to a glittering tabernacle. Inside, he saw the two men. Their faces were bright, their teeth whiter than milk, their eyes shone brighter than the morning star; they were "lofty of stature, of gracious look." Telling him who they were, they said that "God hid them from death." They told him that the place was "The City of the Storehouse of Life," from where the "Bright Waters of Life" emanated. But before Alexander could find out more, or drink of the "Waters of Life," a "chariot of fire" snatched him away—and he found himself back with his troops.

(According to Muslim tradition, the prophet Muhammed was also carried heavenward, a thousand years later, riding his white horse.)

Was the episode of the Cave of the Gods—as the other episodes in the histories of Alexander—pure fiction, mere myth, or perhaps embellished tales based on historical fact?

Was there a Queen Candace, a royal city named Shamar, a world-conqueror named Sesonchusis? In truth, the names meant little to students of antiquity until relatively recently. If these were names of Egyptian royal personages or of a mystical province of Egypt, they were as obscured by time as the monuments were obscured by the encroaching sands; rising above the sands, the pyramids and the Sphinx only broadened the enigma; the hieroglyphic picture-words, undecipherable, only confirmed that there were secrets not to be unlocked. The tales from antiquity, passed on via the Greeks and Romans, dissolved into legends; eventually, they faded into obscurity.

It was only when Napoleon conquered Egypt in 1798, that Europe began to rediscover Egypt. Accompanying Napoleon's troops were groups of serious scholars who began to remove the sands and raise the curtain of forgetfulness. Then, near the village of Rosetta, a stone tablet was found bearing the same inscription in three languages. The key was found to unlock the language and inscriptions of ancient Egypt: its records of Pharaonic feats, the glorification of its gods.

In the 1820s European explorers penetrating southward, into the Sudan, reported the existence of ancient monuments (including sharp-angled pyramids) at a site on the Nile river called Meroe. A Royal Prussian expedition uncovered impressive archaeological remains during excavations in the years 1842–44. Between 1912 and 1914, others uncovered sacred sites; the hieroglyphic inscriptions indicated one of them was called the Sun Temple—perhaps the very place where the spies of Cambyses observed the "Table of the Sun." Further excavations in this century, the piecing together of archaeological finds, and the continued decipherment of the inscriptions, have established that there indeed existed in that land a Nubian kingdom in the first millenium B.C.; it was the biblical Land of Kush.

There indeed was a Queen Candace. The hieroglyphic inscriptions revealed, that at the very beginning of the Nubian kingdom, it was ruled by a wise and benevolent queen. Her name was Candace (Fig. 5). Thereafter, whenever a woman ascended the throne—which was not infrequent—she adopted the name as a symbol of great queenship. And farther south of Meroe, within this kingdom's domain, there was a city named *Sennar*—possibly the *Shamar* referred to in the Alexander tale.

And what about Sesonchusis? It is told in the Ethiopic version of pseudo-Callisthenes, that journeying to (or from) Egypt, Alexander and his men passed by a lake swarming with crocodiles. There, an earlier ruler had built a way to cross the lake. "And behold, there was a building upon the shore of the lake, and above the building was a heathen altar upon which was written: 'I am Kosh, the king of the world, the conqueror who crossed this lake.'"

Fig. 5

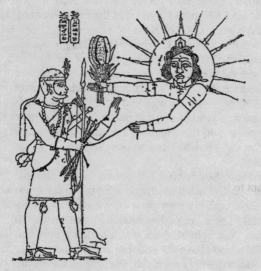

Fig. 6

Who was this world conqueror *Kosh,* namely the king who ruled over Kush or Nubia? In the Greek version of this tale, the conqueror who had commemorated his crossing of the lake—described as part of the waters of the Red Sea—was named Sesonchusis; so Sesonchusis and Kosh were one and the same ruler—a Pharaoh who had ruled both Egypt and Nubia. Nubian monuments depicted such a ruler as he receives from a "Shiny God" the Fruit of Life shaped like date palms (Fig. 6).

Egyptian records do speak of a great Pharaoh who, early in the second millenium B.C., was indeed a world conqueror. His name was Senusert; and he, too, was a devotee of Ammon. Greek historians credited him with the conquest of Libya and Arabia, and significantly also of Ethiopia and all the islands of the Red Sea; of great parts of Asia—penetrating east even farther than the later Persians; and of invading Europe via Asia Minor. Herodotus described the great feats of this Pharaoh, whom he names Sesostris; stating that Sesostris erected memorial pillars wherever he went. "The pillars which he erected," Herodotus wrote, "are still visible." Thus, when Alexander saw the pillar by the lake, it only confirmed what Herodotus had written a century earlier.

Sesonchusis did indeed exist. His Egyptian name meant "He whose births live." For, by virtue of being a Pharaoh of Egypt, he had every right to join the company of the gods, and live forever.

In the search for the Waters of Life or of Eternal Youth, it was important to assert that the search was surely not futile, for others in days past had succeeded in the quest. Moreover, if the waters flow from a Paradise Lost, would not finding those who had been there be a means of learning from them how to get there?

It was with that in mind, that Alexander sought to reach the Immortal Ancestors. Whether he indeed encountered them is not too important: the important fact is that in the centuries preceding the Christian era, Alexander or his historians (or both) believed that the Immortal Ancestors indeed existed—that in days that to them were ancient and olden, mortals could become immortal if the gods so wished.

The authors or editors of the histories of Alexander relate various incidents in which Alexander encountered Sesonchusis; Elijah and Enoch; or just Enoch. The identity of Sesonchusis could only be guessed, and the manner of his translation to Immortality is not described. Not so with Elijah—the companion of Enoch in the Shining Temple, according to one Alexander version.

He was the biblical Prophet who was active in the Kingdom of Israel in the ninth century B.C., during the reign of kings Ahab and Ahaziah. As his adopted name indicated (*Eli-Yah*—"My God is Yahweh"), he was inspired by and stood up for the Hebrew god Yahweh, whose faithful were finding themselves harassed by the followers of the Canaanite god Baal. After a period of seclusion at a secret place near the Jordan River, where he was apparently coached by the Lord, he was given "a mantle of haircloth" of magical powers, and was able to perform miracles. Residing first near the Phoenician town of Sidon, his first reported miracle (as related in I Kings Chapter 17) was the making of a little cooking oil and a spoonful of flour last a widow, who gave him shelter, the rest of her lifetime. Then he prevailed on the Lord to revive her son, after he had died of a violent illness. He could also summon the Fire of God from the skies, which came in handily in the ongoing struggle with the kings and priests who succumbed to pagan temptations.

Of him, the Scriptures say, that he did not die on Earth, for he "went up into Heaven in a whirlwind." According to Jewish traditions, Elijah is still immortal; and to this very day, tradition requires that he be invited to come into Jewish homes on Passover eve. His ascent is described in the Old Testament in great detail. And as reported in II Kings Chapter 2, the event was not a sudden or unexpected occurrence. On the contrary: it was a planned and pre-arranged operation, whose place and time were communicated to Elijah in advance.

The designated place was in the Jordan Valley, on the eastern side of the river—perhaps in the very area where Elijah was ordained as "a Man of God." As he began his last journey to Gilgal—a place commemorating an earlier miracle, as the Bible tells—he had a tough time shaking off his devoted chief disciple Elisha. Along the way, the two Prophets were repeatedly intercepted by disciples, "Sons of Prophets," who kept asking: Is it true that the Lord will take Elijah heavenward today?

Let the biblical narrator tell the story in his own words:

> And it came to pass when the Lord
> would take up Elijah into Heaven by a Whirlwind,
> that Elijah went with Elisha from Gilgal.
> And Elijah said unto Elisha:
> "Tarry here, I pray thee,
> for the Lord has sent me to Beth-El."
> And Elisha said unto him:
> "As the Lord liveth, and by thy life,
> I will not leave thee."
> So they went down to Beth-El.
>
> And the Sons of the Prophets that were at Beth-el
> came forth to Elisha, and said unto him:
> "Knowest thou that the Lord will, this day,
> take the master from above thee?"
> And he said:
> "Yea, I know it too; but keep silent."

Now Elijah admitted to Elisha that his destination was Jericho, by the Jordan River; and he asked his colleague to stay behind. But again Elisha refused and went along with the Prophet; "and so they came to Jericho."

And the Sons of the Prophets that were at Jericho
approached Elisha and said unto him:
"Knowest thou that the Lord will, this day,
take the master from above thee?"
And he said:
"Yea, I know it too; but keep silent."

Foiled thus far in his attempt to proceed alone, Elijah then
asked Elisha to stay behind in Jericho, and to let him proceed
to the river's bank unaccompanied. But Elisha refused, and
would not part from Elijah. Encouraged, "fifty men of the
Sons of the Prophets went along; but they stopped and stood
apart as the two (Elijah and Elisha) reached the Jordan."

And Elijah took his mantle
and rolled it together,
and struck the waters.
And the waters parted hither and thither,
and the two of them crossed over on dry ground.

Once they were across, Elisha asked that Elijah imbue him
with the divine spirit; but before he could get an answer,

As they continued to walk on and to talk,
there appeared a chariot of fire,
and horses of fire, and the two were separated.
And Elijah went up into Heaven,
in a Whirlwind.
And Elisha saw,
and he cried out:
"My father! My father!
The Chariot of Israel and its horsemen!"
And he saw it no more.

Distraught, Elisha sat stunned for a while. Then he saw
that the mantle of Elijah was left behind. Was it by accident or on purpose? Determined to find out, Elisha took the

mantle, and returned to the banks of the Jordan, and called the name of Yahweh, and struck the waters. And lo and behold—"the waters parted hither and thither, and Elisha crossed." And the Sons of the Prophets, the disciples who stood back on the western side of the river in the plain of Jericho, "saw this; and they said: 'the inspiration of Elijah doth rest upon Elisha'; and they came toward him, and prostrated themselves before him."

Incredulous of what they had seen with their own eyes, the fifty disciples wondered whether Elijah was indeed taken heavenward for good. Perchance the Lord's wind had blown him only some distance, and he was thrown upon a mountain or into some ravine? they asked. Over the objections of Elisha, they searched for three days. And when they returned from the futile search, Elisha said: "Did I not say unto you, 'Go not'?" for he well knew the truth: that the Lord of Israel had taken Elijah up in a Chariot of Fire.

The encounter with Enoch, which the histories of Alexander claimed for him, introduced into the Search for Immortality an "Immortal Ancestor" specifically mentioned in the Old and New Testaments alike, the legends of whose ascent to the heavens predated the Bible and were recorded in their own right.

According to the Bible, Enoch was the seventh pre-Diluvial patriarch in the line of Adam through Seth (as distinct from the accursed line of Adam through Cain). He was the great-grandfather of Noah, the hero of the Deluge. The fifth chapter of the Book of Genesis lists the genealogies of these patriarchs, the ages at which their rightful heirs were born, and the ages at which they died. But Enoch was an exception: no mention at all is made of his death. Explaining that "he had walked with the Lord," the Book of Genesis states that at the actual or symbolic age of 365 (the number of days in a solar year), Enoch "was gone" from Earth, "for the Lord had taken him."

Enlarging on the cryptic biblical statement, Jewish commentators often quoted older sources which seemed to describe an actual ascent by Enoch to the heavens, where he was (by some versions) translated into Metatron, the Lord's "Prince of the Countenance" who was stationed right behind the Lord's throne.

According to these legends, as brought together by I. B. Lavner in his *Kol Agadoth Israel [All the Legends of Israel],* when Enoch was summoned to the Lord's abode, a fiery horse was sent for him from the heavens. Enoch was at the time preaching righteousness to the people. When the people saw the fiery horse descending from the skies, they asked Enoch for an explanation. And he told them: "Know ye, that the time has come to leave ye and ascend to Heaven." But as he mounted the horse, the people refused to let him leave, and followed him about for a whole week. "And it was on the seventh day, that a fiery chariot drawn by fiery horses and angels came down, and raised Enoch skyward." While he was soaring up, the Angels of Heaven objected to the Lord: "How comes a man born of a woman to ascend unto the Heavens?" But the Lord pointed out the piety and devotion of Enoch, and opened to him the Gates of Life and of Wisdom, and arrayed him in a magnificent garment and a luminous crown.

As in other instances, cryptic references in the Scriptures often suggest that the ancient editor assumed that his reader was familiar with some other, more detailed writings on the subject at hand. There are even specific mentions of such writings—"The Book of Righteousness," or "The Book of the Wars of Yahweh"—which must have existed, but were entirely lost. In the case of Enoch, the New Testament augments a cryptic statement that Enoch was "translated" by the Lord "that he should not see death" with a mention of a *Testimony of Enoch,* written or dictated by him "before his Translation" to Immortality (Hebrews 11:5). Jude 14, referring to the prophecies of Enoch, is also taken as referring to some actual writings by this patriarch.

Various Christian writings throughout the centuries also contain similar hints or references; and as it turned out, there have in fact circulated since the second century B.C. several versions of a *Book of Enoch*. When the manuscripts were studied in the nineteenth century, scholars concluded that there were basically two sources. The first, identified as *I Enoch* and called the *Ethiopic Book of Enoch,* is an Ethiopic translation of a previous Greek translation of an original work in Hebrew (or Aramaic). The other, identified as *II Enoch,* is a Slavonic translation from an original written in Greek whose full title was *The Book of the Secrets of Enoch.*

Scholars who have studied these versions do not rule out the possibility that both *I Enoch* and *II Enoch* stem from a much earlier original work; and that there indeed could have existed in antiquity a *Book of Enoch. The Apocrypha and Pseudepigrapha of the Old Testament,* which R. H. Charles began to publish in 1913, is still the major English translation of the Books of Enoch and the other early writings which were excluded from the canonized Old and New Testaments.

Written in the first person, *The Book of the Secrets of Enoch* starts with an exact place and time:

> On the first day of the first month of the 365th year I was alone in my house and I rested on my bed and slept. . . . And there appeared to me two men, very tall, such as I have never seen on Earth; and their faces shone like the sun, and their eyes were like burning lamps, and fire came forth from their lips. Their dress had the appearance of feathers, their feet were purple. Their wings were brighter than gold; their hands whiter than snow. They stood at the head of my bed and called me by name.

Because he was asleep when these strangers arrived, Enoch adds for the record that by then he was no longer sleeping; "I saw clearly these men, standing in front of me," he states. He made obeisance to them, and was overtaken by fear. But the two reassured him:

> Be of good cheer, Enoch, be not afraid; the Everlasting God hath sent us to thee and lo, today thou shalt ascend with us into heaven.

They then told Enoch to wake up his family and servants, and order them not to seek him, "till the Lord bring thee back to them." This Enoch did, using the opportunity to instruct his sons in the ways of righteousness. Then the time came to depart:

> It came to pass when I had spoken to my sons, these men summoned me and took me on their wings and placed me on the clouds; and lo, the clouds moved. . . . Going higher I saw the air and (going still) higher I saw the ether; and they placed me in the First Heaven; and they showed me a very great sea, greater than the earthly sea.

Ascending thus unto the heavens upon "clouds that move," Enoch was transported from the First Heaven—where "two hundred angels rule the stars"—to the Second, gloomy Heaven; then to the Third Heaven. There he was shown

> a garden with a goodliness of its appearance; beautiful and fragrant trees and fruits.
> In the midst therein there is a Tree of Life—in that place on which the God rests when he comes into Paradise.

Stunned by the Tree's magnificence, Enoch manages to describe the Tree of Life in the following words: "It is beautiful more than any created thing; on all sides in appearance it is like gold and crimson, transparent as fire." From its root go four streams which pour honey, milk, oil and wine, and they go down from this heavenly Paradise to the Paradise of Eden, making a revolution around Earth. This Third Heaven and its Tree of Life are guarded by three hundred "very glorious" angels. It is in this Third Heaven that the Place of the Righteous, and the Terrible Place where the wicked are tortured, are situated.

Going further up, to the Fourth Heaven, Enoch could see the Luminaries and various wondrous creatures, and the Host of the Lord. In the Fifth Heaven, he saw many "hosts"; in the Sixth, "bands of angels who study the revolutions of the stars." Then he reached the Seventh Heaven, where the greatest angels hurried about and where he saw the Lord— "from afar"—sitting on his throne.

The two winged men and their moving cloud placed Enoch at the limits of the Seventh Heaven, and left; whereupon the Lord sent the archangel Gabriel to fetch Enoch into His Presence.

For thirty-three days, Enoch was instructed in all the wisdoms and all the events of the past and the future; then he was returned to Earth by an awful angel who had a "very cold appearance." In total, he was absent from Earth sixty days. But his return to Earth was only so that he might instruct his sons in the laws and commandments; and thirty days later, he was taken up again unto the heavens—this time, for good.

Written both as a personal testament and as a historic review, the *Ethiopic Book of Enoch,* whose earliest title was probably *The Words of Enoch,* describes his journeys to Heaven as well as to the four corners of Earth. As he traveled north, "toward the north ends of Earth," he "saw there a great and glorious device," the nature of which is not described. And he saw there, as well as at the western ends of Earth, "three portals of heaven open in the heaven" in each place, through which hail and snow, cold and frost blew in.

"And thence I went to the south to the ends of the Earth," and through the portals of Heaven there blow in the dew and rain. And thence he went to see the eastern portals, through which the stars of Heaven pass and run their course.

But the principal mysteries, and secrets of the past and the future, were shown to Enoch as he went to "the middle of the Earth," and to the east and to the west thereof. The "middle of the Earth" was the site of the future Holy Temple in Jerusalem; on his journey east, Enoch reached the Tree of Knowledge; and going west, he was shown the Tree of Life.

On his eastward journey, Enoch passed mountains and deserts, saw water courses flowing from mountain peaks covered by clouds, and snow and ice ("water which flows not"), and trees of diverse fragrances and balsams. Going farther and farther east, he found himself back over mountains bordering the Erythraean Sea (the Sea of Arabia and the Red Sea). Continuing, he passed by Zotiel, the angel guarding the entrance to Paradise, and he "came unto the Garden of Righteousness." There he saw among many wonderful trees the "Tree of Knowledge." It was as high as a fir, its leaves were as of the carob, and its fruit like the clusters of a vine. And the angel who was with him confirmed that indeed it was the very tree whose fruit Adam and Eve had eaten before they were driven out of the Garden of Eden.

On his journey west, Enoch arrived at a "mountain range of fire, which burnt day and night." Beyond it he reached a place encircled by six mountains separated by "deep, rough ravines." A seventh mountain rose in their midst, "resembling the seat of a throne; and fragrant trees encircled the throne. And amongst them was a tree such as I had never smelt . . . and its fruit resembles the dates of a palm."

The angel who accompanied him explained that the middle mountain was the throne "on which the Holy Great One, the Lord of Glory, the Eternal King will sit when He shall come to visit Earth." And as to the tree whose fruits were as the date palms, he said:

As for this fragrant tree, no mortal is permitted to
    touch it till the Great Judgment . . .
Its fruit shall be for food for the elect . . .
Its fragrance shall be in their bones,
And they shall live a long life on Earth.

It was during these journeys that Enoch "saw in those days
how long cords were given to those angels, and they took to
themselves wings, and they went towards the north." And
when Enoch asked what this was all about, the angel who
guided him said: "They have gone off to measure . . . they
shall bring the measures of the righteous to the righteous,
and the ropes of the righteous to the righteous . . . all these
measures shall reveal the secrets of the earth."

Having visited all the secret places on Earth, Enoch's
time had come to take the Journey to Heaven. And, like others after him, he was taken to a "mountain whose summit
reached to Heaven" and to a Land of Darkness:

    And they (the angels) took me to a place in which
those who were there were like flaming fire, and when
they wished, they appeared as men.

    And they brought me to a place of darkness, and
to a mountain the point of whose summit reached to
heaven.

    And I saw the chambers of the luminaries, and the
treasuries of the stars, and of the thunder, in the great
depths, where were a fiery bow and arrows, and their
quiver, and a fiery sword, and all the lightnings.

Whereas, at such a crucial stage, Immortality slipped out
of Alexander's hands because he had searched for it contrary
to his proclaimed destiny—Enoch, as the Pharaohs after
him, was proceeding with divine blessing. Thus, at this crucial moment, he was deemed worthy of proceeding; so "they
(the angels) took me to the Waters of Life."

Continuing, he arrived at the "House of Fire":

And I went in till I drew nigh to a wall which is built of crystals and surrounded by tongues of fire; and it began to affright me.

And I went into the tongues of fire and drew nigh to a large house which was built of crystals; and the walls of the house were like a tesselated floor of crystals, and its groundwork was of crystal. Its ceiling was like the path of the stars and the lightnings, and between them were fiery Cherubim, and their heaven was as water.

A flaming fire surrounded the walls, and its portals blazed with fire.

And I entered into that house, and it was hot as a fire and cold as ice. . . .

And I beheld a vision: behold, there was a second house, greater than the former, and the entire portal stood open before me, and it was built of flames of fire. . . .

And I looked therein and saw a lofty throne: its appearance was as crystal, and the wheels thereof as the shining sun, and there was the appearance of Cherubim.

And from underneath the throne came streams of flaming fire, so that I could not look thereon.

Arriving at the "River of Fire," Enoch was taken aloft. He could see the whole of Earth—"the mouths of all the rivers of Earth . . . and the cornerstones of Earth . . . and the winds on Earth carrying the clouds." Rising higher, he was "where the winds stretch the vaults of Heaven and have their station between Heaven and Earth. I saw the winds of Heaven which turn and bring the circumference of the Sun, and all the stars." Following "the paths of the angels," he reached a point "in the firmament of Heaven above" from which he could see "the end of Earth."

From there, he could view the expanse of the heavens; and he could see "seven stars like great shining mountains"—

"seven mountains of magnificent stones." From wherever he was viewing these celestial bodies, "three were toward the east," where there was "the region of heavenly fire"; there Enoch saw rising and falling "columns of fire"—eruptions of fire "which were beyond measure, alike toward the width and toward the depth." On the other side, three celestial bodies were "toward the south"; there Enoch saw "an abyss, a place which had no firmament of the Heaven above, and no firmly founded Earth below . . . it was a void and awesome place." When he asked the angel who was carrying him aloft for an explanation, he replied: "There the heavens were completed . . . it is the end of Heaven and Earth; it is a prison for the stars and the host of Heaven."

The middle star "reached to Heaven like the throne of God." Having the appearance of alabaster, "and the summit of the throne as of sapphire," the star was "like a flaming fire."

Journeying on in the heavens, Enoch said, "I proceeded to where things were chaotic. And I saw there something horrible." What he saw was "stars of the heaven bound together." And the angel explained to him: "These are of the number of stars of heaven which have transgressed the commandment of the Lord, and are bound here till ten thousand years are consummated."

Concluding his report of the first Journey to Heaven, Enoch said: "And I, Enoch, alone saw the vision, the ends of all things; and no man shall see as I have seen." After being taught at the Heavenly Abode all manner of wisdom, he was returned to Earth to impart teachings to other men. For an unspecified length of time, "Enoch was hidden, and no one of the children of men knew where he was hidden, and where he abode, and what had become of him." But when the Deluge neared, he wrote down his teachings and advised his great-grandson Noah to be righteous and worthy of salvation.

After that, Enoch was once again "raised aloft from among those who dwell on Earth. He was raised aloft on the Chariot of the Spirits, and his 'Name' vanished among them."

III

The Pharaoh's Journey to the Afterlife

The adventures of Alexander and his search for the Immortal Ancestors clearly comprised elements which simulated their experiences: caverns, angels, subterranean fires, fiery horses and Chariots of Fire. But it is equally clear that, in the centuries preceding the Christian era, it was believed (by Alexander or by his historians or by both) that if one wished to attain Immortality, one had to emulate the Egyptian Pharaohs.

Accordingly, Alexander's claim to semi-divine ancestry was evolved from a complicated affair by an Egyptian deity, rather than by simply claiming affinity to a local Greek god. It is an historical fact, not mere legend, that Alexander found it necessary, as soon as he broke through the Persian lines in Asia Minor, not to pursue the Persian enemy, but to go to Egypt; there to seek the answer to his purported divine "roots," and from there to begin the search for the Waters of Life.

Whereas the Hebrews, the Greeks and other peoples in antiquity recounted tales of a unique few who were able to escape a mortal's fate by divine invitation, the ancient Egyptians developed the privilege into a right. Not a universal right, nor a right reserved to the singularly righteous; but a right attendant on the Egyptian king, the Pharaoh, by sole virtue of having sat on the throne of Egypt. The reason for this, according to the traditions of ancient Egypt, was that the first rulers of Egypt were not men but gods.

Egyptian traditions held that in times immemorial "Gods of Heaven" came to Earth from the Celestial Disk (Fig. 7). When Egypt was inundated by waters, "a very great god who came forth (to Earth) in the earliest times" arrived in Egypt and literally raised it from under the waters and mud, by damming the waters of the Nile and undertaking extensive dyking and land reclamation works (it was therefore that Egypt was nicknamed "The Raised Land"). This olden god was named PTAH—"The Developer." He was considered to have been a great scientist, a master engineer and architect, the Chief Craftsman of the gods, who even had a hand in creating and shaping Man. His staff was frequently depicted as a graduated stick—very much like the graduated rod which surveyors employ for field measuring nowadays (Fig. 7).

The Egyptians believed that Ptah eventually retired south, where he could continue to control the waters of the Nile with sluices he had installed in a secret cavern, located at the first cataract of the Nile (the site of today's Aswan Dam). But before leaving Egypt, he built its first hallowed city and named it AN, in honor of the God of the Heavens (the biblical *On*, whom the Greeks called *Heliopolis)*. There, he installed as Egypt's first Divine Ruler his own son RA (so named in honor of the Celestial Globe).

Ra, a great "God of Heaven and Earth," caused a special shrine to be built at An; it housed the *Ben-Ben*—a "secret object" in which Ra had purportedly come down to Earth from the heavens.

In time Ra divided the kingdom between the gods OSIRIS and SETH. But the sharing of the kingdom between the two divine brothers did not work. Seth kept seeking the overthrow and death of his brother Osiris. It took some doing, but finally Seth succeeded in tricking Osiris into entering a coffin, which Seth promptly set to seal and drown. ISIS, the sister and wife of Osiris, managed to find the coffin, which had floated ashore in what is nowadays Lebanon. She hid Osiris as she went to summon the help of other gods who could bring Osiris back to life; but Seth discovered the body

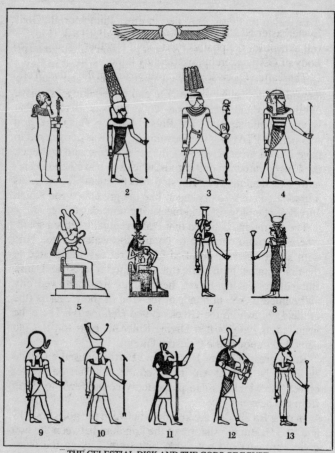

THE CELESTIAL DISK AND THE GODS OF EGYPT

1. Ptah	2. Ra-Amen	3. Thoth	4. Seker
5. Osiris	6. Isis with Horus	7. Nephtys	8. Hathor

The gods with their attributes:

9. Ra/Falcon 10. Horus/Falcon 11. Seth/Sinai Ass. 12. Thoth/Ibis 13. Hathor/Cow

Fig. 7

and cut it to pieces, dispersing them all over the land. Helped by her sister NEPHTYS, Isis managed to retrieve the pieces (all except for the phallus) and to put together the mutilated body of Osiris, thereby resurrecting him.

Thereafter, Osiris lived on, resurrected, in the Other World among the other celestial gods. Of him the sacred writings said:

> He entered the Secret Gates,
> The glory of the Lords of Eternity,
> In step with him who shines in the horizon,
> On the path of Ra.

The place of Osiris on the throne of Egypt was taken over by his son HORUS. When he was born, his mother Isis hid him in the reeds of the river Nile (just as the mother of Moses did, according to the Bible), to keep him out of the reach of Seth. But the boy was stung by a scorpion and died. Quickly, the goddess his mother appealed to THOTH, a god of magical powers, for help. Thoth, who was in the heavens, immediately came down to Earth in Ra's "Barge of Astronomical Years" and helped restore Horus to life.

Growing up, Horus challenged Seth for the throne. The struggle ranged far and wide, the gods pursuing each other in the skies. Horus attacked Seth from a *Nar,* a term which in the ancient Near East meant "Fiery Pillar." Depictions from pre-dynastic times showed this celestial chariot as a long, cylindrical object with a funnel-like tail and a bulkhead from which rays are spewed out, a land of a celestial submarine (Fig. 8). In front the *Nar* had two headlights or "eyes," which according to the Egyptian tales changed color from blue to red.

There were ups and downs in the battles, which lasted several days. Horus shot at Seth, from out of the *Nar,* a specially designed "harpoon," and Seth was hurt, losing his testicles; this only made him madder. In the final battle, over the Sinai peninsula, Seth shot a beam of fire at Horus, and Horus lost

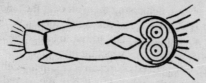

Fig. 8

an "eye." The great gods called a truce and met in council. After some wavering and indecision, the Lord of Earth ruled in favor of giving Egypt to Horus, declaring him the legitimate heir in the Ra-Osiris line of succession. (Thereafter, Horus was usually depicted with the attributes of a falcon, while Seth was shown as an Asiatic deity, symbolized by the ass, the burden animal of the nomads; Fig. 7).

The accession of Horus to the reunited throne of the Two Lands (Upper Egypt and Lower Egypt) remained throughout Egyptian history the point at which kingship was given its perpetual divine connection; for every Pharaoh was deemed a successor of Horus and the occupier of the throne of Osiris.

For unexplained reasons, the rule of Horus was followed by a period of chaos and decline; how long this lasted, no one knows. Finally, circa 3200 B.C., a "dynastic race" arrived in Egypt and a man named Menes ascended the throne of a reunited Egypt. It was then that the gods granted Egypt civilization and what we now call Religion. The kingship that was begun by Menes continued through twenty-six dynasties of Pharaohs until the Persian domination in 525 B.C., and then through Greek and Roman times (when the famed Cleopatra reigned).

When Menes, the first Pharaoh, established the united kingdom, he chose a midpoint in the Nile, just south of Heliopolis, as the place for the capital of the two Egypts. Emulating the works of Ptah, he built *Memphis* on an artificial mound raised above the Nile's waters, and dedicated its temples to Ptah. Memphis remained the political-religious center of Egypt for more than a thousand years.

But circa 2200 B.C. great upheavals befell Egypt, the nature of which is not clear to scholars. Some think that Asiatic invaders overran the country, enslaving the people and disrupting the worship of their gods. Whatever semblance of Egyptian independence remained, it was retained in Upper Egypt—the less accessible regions farther south. When order was restored some 150 years later, political-religious power—the attributes of kingship—flowed from *Thebes,* an old but until then unimposing city in Upper Egypt, on the banks of the Nile.

Its god was called AMEN—"The Hidden One"—the very god Ammon whom Alexander had searched out as his true divine father. As supreme deity, he was worshipped as Amen-Ra, "The Hidden Ra"; and it is not clear whether he was the very same Ra but now somehow unseen or "hidden," or another deity.

The Greeks called Thebes *Diospolis,* "The City of Zeus," for they equated Ammon with their supreme god Zeus. This fact made it easier for Alexander to affiliate himself with Ammon; and it was to Thebes that he rushed after he had received Ammon's favorable oracle at the oasis of Siwa.

There, at Thebes and its precincts (now known as Karnak, Luxor, Dier-el-Bahari), Alexander came upon the extensive shrines and monuments to Ammon—impressive to this very day although they stand empty and in ruins. They were built mostly by Twelfth Dynasty Pharaohs, one of whom was probably the "Sesonchusis" who had searched for the Waters of Life 1,500 years before Alexander. One of the colossal temples was built by Queen Hatshepshut, who was also said to have been a daughter of the god Ammon.

Such tales of divine parentage were not unusual. The Pharaoh's claim to divine status, based on the mere fact of occupying the throne of Osiris, was sometimes augmented by assertions that the ruler was the son or the brother of this or that god or goddess. Scholars consider such statements to have only symbolic meaning; but some Egyptian Pharaohs, such as three kings of the Fifth Dynasty, maintained that they were actually, physically, the sons of the god Ra, begotten by him when he impregnated the wife of the high priest in his own temple.

Other kings attributed their descent from Ra to more sophisticated means. It was claimed that Ra embodied himself in the reigning Pharaoh, through which subterfuge he could then have intercourse with the queen. Thereby, the heir to the throne could claim direct descent of Ra. But apart from such specific claims to be of divine seed, every Pharaoh was theologically deemed to be the incarnation of Horus and thus by extension the son of the god Osiris. Consequently, the Pharaoh was entitled to eternal life in the very same manner experienced by Osiris: to resurrection after death, to an Afterlife.

It was this circle, of gods and god-like Pharaohs, that Alexander longed to join.

The belief was that Ra and the other immortal gods managed to live forever because he kept rejuvenating himself. Accordingly, the Pharaohs bore names meaning, for example, "He Who Repeats Births" and "Repeater of Births." The gods rejuvenated themselves by partaking of divine food and beverage at their abode. Therefore, the king's attainment of an eternal Afterlife called for his joining the gods in their abode, so that he too could partake of their divine sustenances.

The ancient incantations appealed to the gods to share with the deceased king their divine food: "Take ye this king with you, that he may eat of that which ye eat, that he may drink of which ye drink, that he may live on that whereupon ye live." And more specifically, as in a text from the pyramid of King Pepi:

> Give thou sustenance to this King Pepi
> From thy eternal sustenance;
> Thy everlasting beverage.

The departed Pharaoh hoped to draw his everlasting sustenance in the celestial realm of Ra, on the "Imperishable

Star." There, in a mystical "Field of Offerings" or "Field of Life," there grew the "Plant of Life." A text in the pyramid of Pepi I describes him as getting past guards with the appearance of "plumed birds," to be met by the emissaries of Horus. With them

> He traveleth to the Great Lake,
> by which the Great Gods alight.
> These Great Ones of the Imperishable Star
> give unto Pepi the Plant of Life
> whereon they themselves do live,
> so that he may also live thereon.

Egyptian depictions showed the deceased (sometimes with his wife) at this Celestial Paradise, sipping the Waters of Life out of which there grows the Tree of Life with its life-giving fruit, the date palm (Fig. 9).

The celestial destination was the birthplace of Ra, to which he had returned from Earth. There, Ra himself was constantly rejuvenated or "reawakened" by having the Goddess of the Four Jars pour him a certain elixir periodically. It was thus the king's hope to have the same goddess pour

Fig. 9

him too the elixir and "therewith refresh his heart to life." It was in these waters, named "Water of Youth," that Osiris rejuvenated himself; and so it was promised to the departed King Pepi that Horus shall "count for thee a second season of youth"; that he shall "renew thy youth in the waters whose name is 'Water of Youth.'"

Resurrected to Afterlife, even rejuvenated, the Pharaoh attained a paradisical life: "His provision is among the gods; his water is wine, like that of Ra. When Ra eats, he gives to him; when Ra drinks, he gives to him." And in a touch of twentieth century psychotherapy, the text adds: "He sleeps soundly every day . . . he fares better today than yesterday."

The Pharaoh seemed little bothered by the paradox that he had to die first in order to attain Immortality. As supreme ruler of the Two Lands of Egypt, he enjoyed the best possible life on Earth; and the resurrection among the gods was an even more attractive prospect. Besides, it was only his earthly body that was to be embalmed and entombed; for the Egyptians believed that every person possessed a Ba, akin to what we call "soul," which rose heavenward like a bird after death; and a Ka—variably translated Double, Ancestral Spirit, Essence, Personality—through which form the Pharaoh was translated into his Afterlife. Samuel Mercer, in his introduction to the Pyramid Texts, concluded that the Ka stood for the mortal's personification of a god. In other words, the concept implied the existence in Man of a divine element, a celestial or godly Double who could resume life in the Afterlife.

But if Afterlife was possible, it was not easily attained. The departed king had to traverse a long and challenging road, and had to undergo elaborate ceremonial preparations before he could embark on his journey.

The deification of the Pharaoh began with his purification and included embalmment (mummification), so that the dead long would resemble Osiris with all his members tied together. The embalmed Pharaoh was then carried in a funerary procession to a structure topped by a pyramid, in front of which there stood an oval-shaped pillar (Fig. 10).

Within this funerary temple, priestly rites were conducted with a view to achieving for the Pharaoh acceptance at journey's end. The ceremonies, called in the Egyptian funerary texts the "Opening of the Mouth," were supervised by a *Shem* priest—always depicted wearing a leopard skin (Fig. 11). Scholars believe that the ritual was literally what its name implies: the priest, using a bent copper or iron tool, opened the mouth of the mummy or of a statue representing the departed king. But it is clear that the ceremony was primarily symbolic, intended to open for the deceased the "mouth" or Entranceway to the Heavens.

The mummy, by then, was tied up tight in many layers of material and was surmounted by the king's golden death

Fig.10

Fig. 11

mask. Thus, the touching of its mouth (or that of the king's statue) could have been only symbolical. Indeed, the priest intoned not the deceased, but the gods to "open the mouth" so that the Pharaoh could ascend toward eternal life. Special appeals were made to the "Eye" of Horus, lost by him in the battle with Seth, to cause the "opening of the mouth" so that "a path shall be opened for the king among the Shiny Ones, that he may be established among them."

The earthly (and thus by conjecture only temporary) tomb of the Pharaoh—according to the texts and actual archaeological discoveries—had a false door on its eastern side, i.e. the masonry was built there to look like a doorway, but it was actually a solid wall. Purified, with all limbs tied together, "opened of mouth," the Pharaoh was then envisioned as raising himself, shaking off Earth's dust, and exiting by the false door. According to a Pyramid Text which dealt with the resurrection process step by step, the Pharaoh could not pass through the stone wall by himself. "Thou standest at the doors which hold people back," the text said, until "he who is chief of the department"—a divine messenger in charge of this task—"comes out to thee. He lays hold on thy arm, and takes thee to heaven, to thy father."

Aided thus by a divine messenger, the Pharaoh was out of his sealed tomb, through the false door. And the priests broke out in a chant: "The king is on his way to Heaven! The king is on his way to Heaven!"

> The king is on his way to Heaven
> The king is on his way to Heaven
> On the wind, on the wind.
> He is not hindered;
> There is no one by whom he is hindered.
> The king is on his own, son of the gods.
> His bread will come on high, with Ra;
> His offering will come out of the Heavens.
> The king is he "Who Comes Again."

But before the departed king could ascend to Heaven to eat and drink with the gods, he had to undertake an arduous and hazardous Journey. His goal was a land called *Neter-Khert,* "The Land of the Mountain Gods." It was sometimes written pictorially in hieroglyphic by surmounting the symbol for god *(Neter)* ⌐ upon a ferry boat ⌐⌐ ; and indeed, to reach that land, the Pharaoh had to cross a long and winding Lake of Reeds. The marshy waters could be crossed with the aid of a Divine Ferryman, but before he would ferry the Pharaoh over he questioned the king about his origins: What made him think he had the right to cross over? Was he a son of a god or a goddess?

Beyond the lake, past a desert and a chain of mountains, past various guardian gods, lay the *Duat,* a magical "Abode for rising to the Stars," whose location and name have baffled the scholars. Some view it as the Netherworld, the abode of the spirits, where the king must go as Osiris did. Others believe it was an Underworld, and indeed much of its scenes were of a subterranean world of tunnels and caverns with unseen gods, pools of boiling waters, eerie lights, chambers guarded by birds, doors that open by themselves. This magical land was divided into twelve divisions, and was traversed in twelve hours.

The *Duat* was further perplexing, because in spite of its terrestrial nature (it was reached after crossing through a mountain pass) or subterranean aspects, its name was written hieroglyphically with a star and a soaring falcon as its determinatives ⋆ ⌐⌐ or simply with a star within a circle ⊛ , denoting a celestial or heavenly association.

Baffling as it has been, the fact is that the Pyramid Texts, as they followed the Pharaoh's progress through his life, death, resurrection and translation to an Afterlife, considered the human problem to be the inability to fly as the gods do. One text summed up this problem and its solution in two sentences: "Men are buried, the gods fly up. Cause this king to fly to Heaven, (to be) among his brothers the gods." A text inscribed in the pyramid of King Teti expressed the Pharaoh's hope and appeal to the gods in these words:

> Men fall,
> They have no Name.
> Seize thou king Teti by his arms,
> Take thou king Teti to the sky,
> That he die not on Earth among men.

And so it was incumbent upon the king to reach the "Hidden Place," and go through its subterranean labyrinths until he could find there a god who carries the emblem of the Tree of Life, and a god who is the "Herald of Heaven." They will open for him secret gates, and lead him to the Eye of Horus, a Celestial Ladder into which he would step—an object which can change hues to blue and red as it is "powered." And then, himself turned into the Falcon-god, he would soar skyward to the eternal Afterlife on the Imperishable Star. There, Ra himself would welcome him:

> The Gates of Heaven are opened for thee;
> The doors of the Cool Place are opened for thee.
> Thou shalt find Ra standing, waiting for thee.
> He will take your hand,
> He will take thee to the Dual Shrine of Heaven;
> He will place thee on the throne of Osiris . . .
> Thou shalt stand supported, equipped as a god . . .
> Among the Eternals, on the Imperishable Star.

Much of what is known today on the subject comes from the Pyramid Texts—thousands of verses combined into hundreds of Utterances, that were discovered embossed or painted (in the hieroglyphic writing of ancient Egypt) on the walls, passages and galleries of the pyramids of five Pharaohs (Unas, Teti, Pepi I, Merenra and Pepi II) who ruled Egypt from circa 2350 B.C. to 2180 B.C. These texts were sorted out and numbered by Kurt Sethe in his masterful *Die altaegyptischen Pyramidentexte,* which has remained the major reference source together with the English counterpart, *The Pyramid Texts* by Samuel A. B. Mercer.

The thousands of verses that make up the Pyramid Texts seem to be just a collection of repetitious, unconnected incantations, appeals to the gods or exaltations of the king. To make some sense of the material, scholars have developed theories about shifting theologies in ancient Egypt, a conflict and then a merger between a "Solar Religion" and a "Sky Religion," a priesthood of Ra and one of Osiris, and so on, pointing out that we deal with material that has been accumulated over millenia.

To scholars who view the mass of verses as expressions of primitive mythologies, figments of the imagination of people who cowered in fear as the wind howled and the thunder roared and called these phenomena "gods"—the verses remain as puzzling and confusing as ever. But these verses, all scholars agree, were extracted by the ancient scribes from older and apparently well-organized, cohesive and comprehensible scriptures.

Later inscriptions on sarcophagi and coffins, as well as on papyrus (the latter usually accompanied by illustrations) indeed show that the verses, Utterances and Chapters (bearing such names as "Chapter of those who ascend") were copied from "Books of the Dead," which bore such titles as "That Which Is in the *Duat,*" "The Book of the Gates," "The Book of the Two Ways." Scholars believe that these "books" in turn were versions of two earlier basic works: olden writings that dealt with the celestial journey of Ra, and a later source which stressed the blissful Afterlife of those who join Osiris resurrected: the food, the beverage, the conjugal joys in a heavenly abode. (Verses of this version were even inscribed on talismans, to achieve for their wearer "union with women by day or night" and the "desire of women" at all times.)

The scholarly theories, however, leave unexplained the magical aspects of the information offered by these texts. Bafflingly, an Eye of Horus is an object existing independently of him—an object into whose insides the king can enter, and which can change hues to blue and red as it is "powered." There exist self-propelled ferries, doors that open by themselves, unseen gods whose faces radiate a glow. In the Underworld, supposedly inhabited by spirits only, "bridge girders" and "copper cables" are featured. And the most baffling aspect of all: Why, if the Pharaoh's transfiguration takes him to the Underworld, do the texts claim that "the king is on his way to *Heaven*"?

Throughout, the verses indicate that the king is following the route of the gods, that he is crossing a lake the way a god had crossed it before, that he uses a barque as the god Ra had done, that he ascends "equipped as a god" as Osiris was, and so on and on. And the question arises: What if these texts were not primitive fantasies—mythology—but accounts of a simulated journey, wherein the deceased Pharaoh emulated what the gods had actually done? What if the texts, substituting the name of the king for that of a god, were copies of some much earlier scriptures that dealt not with the journeys of the Pharaohs, but with the journeys of the gods?

One of the early leading Egyptologists, Gaston Maspero (*L'Archéologie égyptienne* and other works), judging by grammatical form and other evidence, suggested that the Pyramid Texts originated at the very beginning of Egyptian civilization, perhaps even before they were written down hieroglyphically. J. H. Breasted has more recently concluded (*Development of Religion and Thought in Ancient Egypt*) that "such older material existed, whether we possess it or not." He found in the texts information on the conditions of civilization and events which enhances the veracity of the texts as conveyors of factual information and not of fantasy. "To one of quick imagination," he says, "they abound in pictures from the long-vanished world of which they are a reflection."

Taken together, the texts and later illustrations describe a journey to a realm that begins above ground, that leads underground, and that ends with an opening to the skies

through which the gods—and the kings emulating them—
were launched heavenward (Fig. 12). Thus the hieroglyphic
connotation combining a subterranean place with a celestial
function.

Have the Pharaohs, journeying from their tombs to the
Afterlife, actually taken this Route to Heaven? Even the ancient Egyptians claimed the journey not for the mummified
corpse, but for the Ka (Double) of the departed king. But
they have envisioned this Double as re-enacting actual progress through actual places.

What, then, if the texts reflect a world which had indeed
existed—what if the Pharaoh's Journey to Immortality, even
if only by emulation, indeed followed step by step actual
journeys undertaken in prehistoric times?

Let us follow in these footsteps; let us take the Route of
the Gods.

Fig. 12

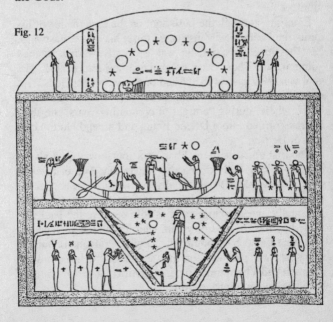

IV

The Stairway to Heaven

Let us imagine ourselves in the Pharaoh's magnificent funerary temple. Having mummified and prepared the Pharaoh for his Journey, the *Shem* priests now intone the gods to open for the king a path and a gateway. The divine messenger has arrived on the other side of the false door, ready to take the Pharaoh through the stone wall and launch him on his journey.

Emerging through the false door on the eastern side of his tomb, the Pharaoh was instructed to set his course eastward. Lest he misunderstand, he was explicitly warned against going west: "Those who go thither, they return not!" His goal was the *Duat,* in the "Land of the Mountain Gods." He was to enter there "The Great House of Two . . . the House of Fire"; where, during "a night of computing years," he shall be transformed into a Divine Being and ascend "to the east side of Heaven."

The first obstacle in the Pharaoh's course was the Lake of Reeds—a long body of marshy waters made up of a series of adjoining lakes. Symbolically, he had the blessing of his guardian god to cross the lake by parting its waters (Fig. 13); physically, the crossing was possible because the lake was served by the Divine Ferryman, who ferried the gods across in a boat made by Khnum, the Divine Craftsman. But the Ferryman was stationed on the far side of the lake, and the Pharaoh had a hard time convincing him that he was entitled to be fetched and ferried over.

The Ferryman questioned the Pharaoh about his origins.

Fig. 13

Was he the son of a god or goddess? Was he listed in the "Register of the Two Great Gods"? The Pharaoh explained his claims to being of "divine seed," and gave assurances of his righteousness. In some cases it worked. In other instances the Pharaoh had to appeal to Ra or to Thoth to get him across; in which instances, the boat and its oars or rudder came alive with uncanny forces: the ferryboat began to move by itself, the steering-oar grasped by the king directed itself. All, in short, became self-propelled. One way or another, the Pharaoh managed to cross the lake and be on his way toward "The Two That Bring Closer the Heavens":

> He descends into the boat, like Ra,
>   on the shores of the Winding Watercourse.
> The king rows in the *Hanbu*-boat;
> He takes the helm toward the
> Plain of "The Two That Bring Closer, the Heavens,"
>   in the land beginning from the Lake of Reeds.

The Lake of Reeds was situated at the eastern end of the domain of Horus. Beyond lay the territories of his adversary Seth, the "lands of Asia." As would be expected on such a sensitive boundary, the king discovers that the lake's eastern shore is patrolled by four "Crossing guards, the wearers of side locks." The way these guards wore their hair was truly their most conspicuous feature. "Black as coal," it was "arranged in curls on their foreheads, at their temples and at the back of their heads, with braids in the center of their heads."

Combining diplomacy with firmness, the king again proclaimed his divine origins, claiming he was summoned by "my father Ra." One Pharaoh is reported to have used threats: "Delay my crossing, and I will pluck out your locks as lotus flowers are plucked in the lotus pond!" Another had some of the gods come to his assistance. One way or another, the Pharaoh managed to proceed.

The king has now left the lands of Horus. The eastward place which he seeks to reach—though under the aegis of Ra—is "in the region of Seth." His goal is a mountainous area, the Mountains of the East (Fig. 14). His course is set toward a pass between two mountains, "the two mountains which stand in awe of Seth." But first he has to traverse an arid and barren area, a kind of no-god's land between the domains of Horus and Seth. Just as the pace and urgency of the Utterances increase, for the king is getting closer to the Hidden Place where the Doors of Heaven are located, he is challenged again by guards. "Where goest thou?" they demand to know.

The king's sponsors answer for him: "The king goes to Heaven, to possess life and joy; that the king may see his father, that the king may see Ra." As the guards contemplate the request, the king himself pleads with them: "Open the frontier . . . incline its barrier . . . let me pass as the gods pass through!"

Having come from Egypt, from the domain of Horus, the king and his sponsors recognize the need for prudence. Many Utterances and verses are employed to present the king as neutral in the feud between the gods. The king is introduced both as "born of Horus, he at whose name the Earth quakes," and as "conceived by Seth, he at whose name Heaven trembles." The king stresses not only his affinity to Ra, but declares that he proceeds "in the service of Ra"; producing thereby a *laissez-passer* from higher authority. With shrewd evenhandedness, the texts point out to the two gods their own self-interest in the king's continued journey, for Ra would surely appreciate their aid to one who comes in his service.

Finally, the guards of the Land of Seth let the king proceed toward a mountain pass. The king's sponsors make sure that he realizes the import of the moment:

> Thou are now on the way to the high places
> In the land of Seth.
> In the land of Seth
> Thou will be set on the high places,
> On that high Tree of the Eastern Sky
> On which the gods sit.

The king has arrived at the *Duat*.

Fig. 14

Fig. 15

The *Duat* was conceived as a completely enclosed Circle of the Gods (see Fig. 15), at the head-point of which there was an opening to the skies (symbolized by the goddess Nut) through which the Imperishable Star (symbolized by the Celestial Disk) could be reached. Other sources suggested in reality a more oblong or oval valley, enclosed by mountains. A river which divided into many streams flowed through this land, but it was hardly navigable and most of the time Ra's barge had to be towed, or moved by its own power as a "boat of earth," as a sled.

The *Duat* was divided into twelve divisions, variably

described as fields, plains, walled circles, caverns or halls, beginning above ground and continuing underground. It took the departed king twelve hours to journey through this enchanted and awesome realm; this he could achieve, because Ra had put at his disposal his magical barge or sled, in which the king traveled aided and protected by his sponsoring gods.

There were seven gaps or passes in the mountains that enclosed the *Duat,* and two of them were in the mountains on the east side of Egypt 𓂝𓏴𓈖 (i.e. in the mountains on the west of the *Duat)*, which were called "The Horizon" or "The Horn" of "The Hidden Place." The pass through which Ra had traveled was 220 *atru* (some twenty-seven miles) long, and followed the course of a stream; the stream, however, ran dry and Ra's barge had to be towed. The pass was guarded and had fortifications "whose doors were strong."

The Pharaoh, as some papyri indicate, took the course leading through the second, shorter pass (only some fifteen miles long). The papyrus drawings show him upon the barge or sled of Ra, passing between two mountain peaks on each of which there is stationed a company of twelve guardian gods. The texts describe a "Lake of Boiling Waters" nearby—waters which, despite their fiery nature are cool to the touch. A fire burns below the ground. The place has a strong bituminous or "natron" stench which drives away the birds. Yet not too far away, there is depicted an oasis with shrubs or low trees around it.

Once across the pass, the king encounters other companies of gods. "Come in peace," they say. He has arrived at the second division.

It is called, after the stream that runs through it, *Ur-nes* (a name which some scholars equate with *Uranus,* the Greek god of the skies). Measuring some fifteen by thirty-nine miles, it is inhabited by people with long hair, who eat the flesh of their asses and depend on the gods for water and sustenance, for the place is arid and the streams are mostly dry. Even Ra's barge turns here into a "boat of earth." It is a domain associated with the Moon god, and with Hathor, the Goddess of Turquoise.

Aided by the gods, the king passes safely through the second division and in the Third Hour arrives at *Net-Asar,* "The Stream of Osiris." Similar in size to the second division, this third division is inhabited by "The Fighters." It is there that the four gods, who are in charge of the four cardinal points of the compass, are stationed.

The pictorial depictions which accompanied the hieroglyphic texts surprisingly showed the Stream of Osiris as meandering its way from an agricultural area, through a chain of mountains, to where the stream divided into tributaries. There, watched over by the legendary Phoenix birds, the *Stairway to Heaven* was situated; there, the Celestial Boat of Ra was depicted as sitting atop a mountain, or rising heavenward upon streams of fire (Fig. 16).

Here, the pace of prayers and Utterances increases again. The king invokes the "magical protectors," that "this man of Earth may enter the *Neter-Khert*" unmolested. The king is nearing the heart of the Duat, he is near the *Amen-Ta,* the "Hidden Place."

It was there that Osiris himself had risen to the Eternal Afterlife. It was there that the "Two That Bring Closer the Heaven" stood out "yonder against the sky," as two magical trees. The king offers a prayer to Osiris (the Chapter's title in the Book of the Dead is "Chapter of Making His *Name* in the *Neter-Khert* Granted"):

> May be given to me my *Name*
>   in the Great House of Two;
> May in the House of Fire
>   my Name be granted.
> In the night of computing years,
>   and of telling the months,
>   may I be a Divine Being,
>   may I sit at the east side of Heaven.
> Let the god advance me from behind;
> Everlasting is his *Name*.

Fig. 16

The king is within sight of the "Mountain of Light."
He has reached the STAIRWAY TO HEAVEN.

The Pyramid Texts said of the place that it was "the stairway in order to reach the heights." Its stairs were described as "the stairs to the sky, which are laid out for the king, that he may ascend thereon to the heavens." The hieroglyphic pictograph for the Stairway to Heaven was sometimes a single stairway ◿ (which was also cast in gold and worn as a charm), or more often a double stairway △, as a step pyramid. This Stairway to Heaven was constructed by the gods of the city of An—the location of the principal temple of Ra—so that they, the gods, could be "united with the Above."

The king's goal is the Celestial Ladder, an Ascender which would actually carry him aloft. But to reach it in the House of Fire, the Great House of Two, he must enter the *Amen-Ta,* the Hidden Land of Seker, God of the Wilderness.

It is a domain described as a fortified circle. It is the subterranean Land of Darkness, reachable by entering into a mountain and going down spiraling hidden paths protected by secret doors. It is the fourth division of the *Duat* which the king must now enter; but the mountain entrance is protected by two walls and the passage between them is swept by flames and manned by guarding gods.

When Ra himself had arrived at this entrance to the Hidden Place, "he performed the designs"—followed the procedures—"of the gods who are therein by means of his voice, without seeing them." But can the king's voice alone achieve for him admission? The texts remind the challenger that only "he who knoweth the plan of the hidden shaftways which are in the Land of Seker," shall have the ability to journey through the Place of Underground Passages and eat the bread of the gods.

Once again the king offers his credentials. "I am the Bull, a son of the ancestors of Osiris," he announces. Then the gods who sponsor him, pronounce in his behalf the crucial words for admission:

> Admittance is not refused thee
> At the gate of the *Duat;*
> The folding doors of the Mountain of Light
> Are opened to thee;
> The bolts open to thee of themselves.
> Thou treadest the Hall of the Two Truths;
> The god who is in it greets thee.

The right formula or password having thus been pronounced, a god named Sa uttered a command; at his word, the flames ceased, the guards withdrew, the doors opened automatically, and the Pharaoh was admitted into the subterranean world.

"The mouth of the earth opens for thee, the eastern door of heaven is open for thee," the gods of the *Duat* announce to the king. He is reassured that though he enters the mouth of the earth, it is indeed the Gateway to Heaven, the coveted eastern door.

The journey in the fourth and following Hours leads the king through caverns and tunnels where gods of diverse functions are sometimes seen, sometimes only heard. There are underground canals, on which gods move about in soundless barques. There are eerie lights, phosphorous waters, torches that light the way. Mystified and terrified, the king moves on, toward "the pillars that reach the Heaven."

The gods seen along the way are mostly organized in groups of twelve, and bear such epithets as "Gods of the Mountain," "Gods of the Mountain of the Hidden Land," or "The Holders of the Time of Life in the Hidden Land." The drawings that accompanied some of the ancient texts provide identification of these gods through the different scepters held by them, their particular headgear, or by depicting their animal attributes— hawk-headed, jackal-headed, lion-headed. Serpents also make an appearance, representing subterranean guards or servants of the gods in the Hidden Land.

The texts and the ancient illustrations suggest that the king has entered a circular underground complex, within which a vast tunnel first spirals down and then up. The depictions, presented in a cross-section fashion, show a gradually sloping tunnel some forty feet high, with a smooth ceiling and a smooth floor, both made of some solid material two to three feet thick. The tunnel is partitioned into three levels, and the king moves within the middle level or corridor. The upper and lower levels are occupied by gods, serpents and structures of diverse functions.

The king's sled, pulled by four gods, begins its journey by gliding silently along the middle corridor; only a beam emitted from the vehicle's bow lights the way. But soon the passage is blocked by a sharply slanting partition, and the king must get off and continue on foot.

The partition, as the cross-section depictions show, is one wall of a shaft that cuts across the three tunnel levels (which slope at about 15°) at a sharper angle of some 40°. It apparently begins above the tunnel, perhaps at ground level or somewhere higher within the mountain; it seems to end as it reaches the floor of the lowest, third level. It is called *Re-Stau,* "The Path of the Hidden Doors"; and at the first and second levels, it is indeed provided with chambers that look like air-locks. These chambers enable Seker and other "hidden gods" to pass through, though "the door has no leaves." The king, who has left his sled, mysteriously passes through this slanting wall simply by virtue of the command of some god, whose voice had activated the air-lock. He is greeted on the other side by representatives of Horus and Thoth, and is passed along from god to god. (Fig. 17.)

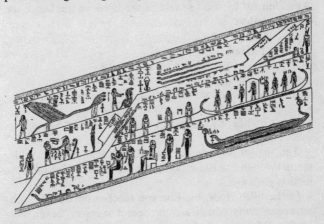

Fig. 17

On his way down, the king sees "faceless gods"—gods whose faces cannot be seen. Offended or simply curious, he pleads with them:

> Uncover your faces,
>> take off your head coverings,
>> when ye meet me;
> For, behold, I [too] am a mighty god
>> come to be among you.

But they do not heed his plea to show their faces; and the texts explain that even they, "these hidden beings, neither see nor look upon" their own chief, the god Seker "when he is in this form himself, when he is inside his abode in the earth."

Spiraling his way down, the king passes through a door and finds himself on the third, lowest level. He enters an antechamber which bears the emblem of the Celestial Disk, and is greeted by the god who is "The Messenger of Heaven" and a goddess who wears the feathered emblem of Shu, "He who rested the sky upon the Stairway to Heaven" (Fig. 18). As called for by the formula in the Book of the Dead, the king proclaims:

> Hail,
>> two children of Shu!
> Hail,
>> children of the Place Of The Horizon . . .
> May I ascend?
> May I journey forth like Osiris?

The answer must be positive, for the king is admitted by them, through a massive door, into the shafts which only the hidden gods use.

In the Fifth Hour, the Pharaoh reaches the deepest subterranean parts which are the secret ways of Seker. Following shafts that incline up, over and down, the Pharaoh cannot see Seker; but the cross-section drawings depict the

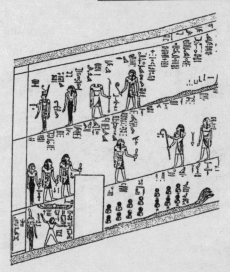

Fig. 18

god as a hawk-headed person, standing upon a serpent and holding two wings within a completely enclosed oval structure deep underground, guarded by two sphinxes. Though the king cannot see this chamber, he hears coming from it "a mighty noise, like that heard in the heights of the heavens when they are disturbed by a storm." From the sealed chamber there flows a subterranean pool whose "waters are like fire." Chamber and pool alike are in turn enclosed by a bunkerlike structure, with a compartmentalized air-lock on the left side and a huge door on the right side. As further protection, a mound of soil is piled up atop the sealed chamber. The mound is topped by a goddess, whose head only is seen, protruding into the descending corridor. A beetle symbol (meaning "to roll, to come into being") connects the head of the goddess with a conical chamber or object in the uppermost corridor (Fig. 19); two birds are perched upon it.

The texts and symbols inform us that, though Seker was

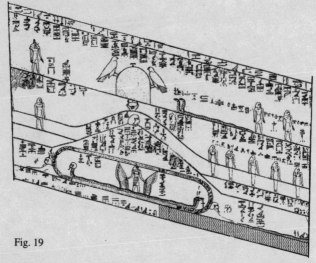

Fig. 19

hidden, his presence could be made known even in the darkness, because of a glowing "through the head and eyes of the great god, whose flesh radiates forth light." The triple arrangement—goddess, beetle *(Kheper)* and conical object or chamber—apparently served to enable the hidden god to be informed of what goes on outside his hermetically sealed chamber. The hieroglyphic text adjoining the beetle symbol states: "Behold Kheper who, immediately the (boat?) is towed to the top of this circle, connects himself with the ways of the *Duat.* When this god standeth on the head of the goddess, he speaks words to Seker every day."

The passage by the Pharaoh over the hidden chamber of Seker and by the setup through which Seker was informed of such passage, was deemed a crucial phase in his progress. The Egyptians were not the only ones in antiquity who believed that each departed person faced a moment of judgment, a spot where their deeds or hearts would be weighed and evaluated and their soul or Double either condemned to the Fiery Waters of Hell, or blessed to enjoy the cool and lifegiving waters of Paradise. By ancient accounts, here was such a Moment of Truth for the Pharaoh.

Speaking for the Lord of the *Duat,* the goddess whose head only was seen announced to the Pharaoh the favorable decision: "Come in peace to the *Duat* . . . advance in thy boat on the road which is in the earth." Naming herself Ament (the female Hidden One), she added: "Ament calls to thee, so that thou mayest go forward in the sky, as the Great One who is in the Horizon."

Passing the test, not dying a second time, the king was born again. The way now led by a row of gods whose task it was to punish the condemned; but the king proceeds unharmed. He rejoins his boat or sled; it is accompanied by a procession of gods; one of them holds the emblem of the Tree of Life (Fig. 20).

The king has been found worthy of Afterlife.

Leaving the zone of Seker, the king enters the sixth division, associated with Osiris. (In versions of the Book of the Gates,

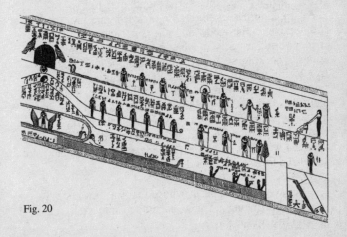

Fig. 20

it was in this Sixth Hour that Osiris judged the departed.) Jackal-headed gods "Who Open the Ways" invite the king to take a refreshing dip in the subterranean pool or Lake of Life, as the Great God himself had done when he passed here before. Other gods, "humming as bees," reside in cubicles whose doors fly open by themselves as the king moves by. As he progresses, the epithets of the gods assume more technical aspects. There are the twelve gods "who hold the rope in the *Duat,*" and the twelve "who hold the measuring cord."

The sixth division is occupied by a series of chambers set close together. A curving path is called "The Secret Path of the Hidden Place." The king's boat is towed by gods clad in leopard skins, just as the *Shem* priests who performed the Opening of the Mouth ceremonies were clad.

Is the king nearing the Opening or Mouth of the Mountain? In the Book of the Dead, the chapters indeed now bear such titles as "The chapter of sniffing the air and of getting power." His vehicle is now "endowed with magical powers . . . he journeyeth where there is no stream and where there are none to tow him; he performeth this by words of power" which proceed from the mouth of a god.

As the king passes through a guarded gate into the seventh division, the gods and the surroundings lose their "underworld" aspects and begin to assume celestial affiliations. The king encounters the falcon-headed god *Heru-Her-Khent,* whose hieroglyphic name included the stairway symbol and who wore on his head the Celestial Disk emblem. His task is "to send the star-gods on their way and to make the constellation-goddesses go on their way." These were a group of twelve gods and twelve goddesses who were depicted with star emblems. The incantations to them were addressed to "the starry gods"—

who are divine in flesh, whose magical powers have come into being . . . who are united into your stars, who rise up for Ra . . .

Let your stars guide his two hands so that he may journey to the Hidden Place in peace.

In this division, there are also present two companies of gods associated with the *Ben-ben,* the mysterious object of Ra that was kept at his temple in the city of An (Heliopolis). They "are those who possess the mystery," guarding it inside the *Het-Benben* (The Ben-ben House); and eight who guard outside but also "enter unto the Hidden Object." Here there are also nine objects, set up in a row, representing the symbol *Shem* which hieroglyphically meant "Follower."

The king has indeed arrived in parts of the *Duat* associated with An, after whom Heliopolis was named. In the Ninth Hour, he sees the resting place of the twelve "Divine Rowers of the Boat of Ra," they who operate Ra's celestial "Boat of Millions of Years." In the Tenth Hour, passing through a gate, the king enters a place astir with activity. The task of the gods there is to provide Flame and Fire to the boat of Ra. One of the gods is called "Captain of the gods of the boat." Two others are those "Who order the course of the stars." They and other gods are depicted with one, two or three star symbols, as though showing some rank associated with the heavens.

Passing from the tenth to the eleventh division, the affinity to the heavens rapidly increases. Gods bear the Celestial Disk and star emblems. There are eight goddesses with star emblems "who have come from the abode of Ra." The king sees the "Star Lady" and the "Star Lord," and gods whose task it is to provide "power for emerging" from the *Duat,* "to make the Object of Ra advance to the Hidden House in the Upper Heavens."

In this place there are also gods and goddesses whose task it is to equip the king for a celestial trip "over the sky." Together with some gods he is made to enter a "serpent" inside which he is to "shed the skin" and emerge "in the form of a rejuvenated Ra." Some of the terms here employed in the texts are still not understood, but the process is clearly explained: the king, having entered dressed as he came, emerges as a falcon, "equipped as a god": the king "lays down on the ground the *Mshdt*-garment"; he puts on his back the "Mark-garment"; he "takes his divine *Shuh*-vestment" and he puts on "the collar of beloved Horus" which is like "a collar on the neck of Ra." Having done all that, "the king has established himself there as a god, like them." And he tells the god who is with him: "If thou goest to Heaven, so will the king go to Heaven."

The illustrations in the ancient texts depict here a group of gods dressed in unusual garb, like tightly fitting overalls adorned with circular collar bands (Fig. 21).

They are led or directed by a god with the emblem of the Celestial Disk upon his head, who stands with outstretched arms between the wings of a serpent with four human legs. Against a starry background, the god and the serpent face another serpent which, though wingless, clearly flies as it carries aloft a seated Osiris. (Fig. 22).

Having been properly equipped, the king is led to an opening in the center of a semi-circular wall. He passes the hidden door. Now he moves within a tunnel which is "1300 cubits long" called "Dawn at the End." He reaches a vestibule; the emblems of the Winged Disk are seen everywhere. He encounters goddesses "who shed light upon the road of Ra" and a magical scepter representing "Seth, the Watcher."

The gods explain to the awed king:

> This cavern is the broad hall of Osiris
> Wherein the wind is brought;
> The north wind, refreshing,
> Will raise thee, O king, as Osiris.

It is now the twelfth division, the final Hour of the king's subterranean journey. It is "the uttermost limit of the thick darkness." The point which he has reached is named "Mountain of the Ascent of Ra." The king looks up and is startled:

Fig. 21

Fig. 22

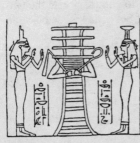

Fig. 23   a   b

the celestial boat of Ra looms in front of his eyes, in all its awesome majesty.

He has reached an object which is called "The Ascender to the Sky." Some texts suggest that Ra himself prepared the Ascender for the king, "that the king may ascend upon it to the heavens"; other texts say that the Ascender was made or set up by several other gods. It is "the Ascender which had carried Seth" heavenward. Osiris could not reach the Firmament of Heaven except by means of such an Ascender; thus the king too requires it in order to be translated, as Osiris, to eternal life.

The Ascender or Divine Ladder was not a common ladder. It was bound together by copper cables; "its sinews (like those) of the Bull of Heaven." The "uprights at its sides" were covered over tightly with a kind of "skin"; its rungs were "*Shesha*-hewn" (meaning unknown); and "a great support (was) placed under it by He Who binds."

Illustrations to the Book of the Dead showed such a Divine Ladder—sometimes with the *Ankh* ("Life") sign ☥ symbolically reaching toward the Celestial Disk in the heavens—in the shape of a high tower with a superstructure (Fig. 23a, b). In stylized form, the tower by itself was written hieroglyphically 𓊽 *("Ded")* and meant "Everlastingness." It was a symbol most closely associated with Osiris, for a pair of such pillars 𓊽𓊽 were said to have been erected in front of his principal temple at Abydos, to commemorate the two objects which stood in the Land of Seker and made possible the ascent of Osiris heavenward.

A long Utterance in the Pyramid Texts is both a hymn to the Ascender—the "Divine Ladder"—and a prayer for its granting to the king Pepi:

> Greeting to thee, divine Ascender;
> Greeting to thee, Ascender of Seth.
> Stand thou upright, Ascender of god;
> Stand upright, Ascender of Seth;

Stand upright, Ascender of Horus
    whereby Osiris came forth into Heaven . . .
Lord of the Ascender . . .
To whom shalt thou give the Ladder of god?
To whom shall thou give the Ladder of Seth,
That Pepi may ascend to Heaven on it,
    to do service as a courtier of Ra?
Let also the Ladder of god be given to Pepi,
Let the Ladder of Seth be given to Pepi
    that Pepi may ascend to Heaven on it.

The Ascender was operated by four falcon-men, "Children of Horus" the Falcon-god, who were "the sailors of the boat of Ra." They were "four youths," who were "Children of the Sky." It is they "who come from the eastern side of the sky . . . who prepare the two floats for the king, that the king may thereby go to the horizon, to Ra." It is they who "join together"—assemble, prepare—the Ascender for the king: "They bring the Ascender . . . they set up the Ascender . . . they raise up the Ascender for the king . . . that he might ascend to Heaven on it."

The king offers a prayer:

May my "Name" to me be given
    in the Great House of Two;
May my "Name" be called
    in the House of Fire,
    in the night of Computing Years.

Some illustrations show the king being granted a *Ded*—"Everlastingness." Blessed by Isis and Nephtys, he is led by a falcon-god to a rocket-like *Ded,* equipped with fins (Fig. 24).

The king's prayer to be given Everlastingness, a "Name," a Divine Ladder, has been granted. He is about to begin his actual ascent to the Heavens.

*   *   *

Fig. 24

Though he requires only one Divine Ladder for himself, not one but two Ascenders are raised together. Both the "Eye of Ra" and the "Eye of Horus" are prepared and put into position, one on the "wing of Thoth" and the other on the "wing of Seth." To the puzzled king, the gods explain that the second boat is for the "son of Aten," a god descended of the Winged Disk—perhaps the god to whom the king had spoken in the "equipping chamber":

> The Eye of Horus is mounted
> Upon the wing of Seth.
> The cables are tied,
>   the boats are assembled,
> That the son of the Aten
>   be not without a boat.
> The king is with the son of Aten;
> He is not without a boat.

"Equipped as a god," the king is assisted by two goddesses "who seize his cables" to step into the Eye of Horus. The term "Eye" (of Horus, of Ra) which has gradually replaced the term Ascender or Ladder, now is being increasingly displaced by the term "boat." The "Eye" or "boat" into which the king steps in is 770 cubits (circa 1000 feet) long. A god who is in charge of the boat sits at its bow. He is instructed: "Take this king with thee in the cabin of thy boat."

As the king "steps down into the perch"—a term denoting an elevated resting place, especially of birds—he can see the face of the god who is in the cabin, "for the face of the god is open." The king "takes a seat in the divine boat" between two gods; the seat is called "Truth which makes alive." Two "horns" protrude from the king's head (or helmet); "he attaches to himself that which went forth from the head of Horus." He is plugged-in for action.

The texts dealing with the Journey to the Afterlife by King Pepi I describe the moment: "Pepi is arrayed in the apparel of Horus, and in the dress of Thoth; Isis is before him and Nephtys is behind him; Ap-uat who is Opener of the Ways hath opened a way unto him; Shu the Sky Bearer hath lifted him up; the gods of An make him ascend the Stairway and set him before the Firmament of the Heaven; Nut the sky goddess extends her hand to him."

The magical moment has arrived; there are only two more doors to be opened, and the king—as Ra and Osiris had done before—will emerge triumphantly from the *Duat* and his boat will float on the Celestial Waters. The king says a silent prayer: "O Lofty one . . . thou Door of Heaven: the king has come to thee; cause this door to be opened for him." The "two *Ded* pillars are standing" upright, motionless.

And suddenly "the double doors of heaven are open!"

The texts break out in ecstatic pronouncements:

> The Door to Heaven is open!
> The Door of Earth is open!
> The aperture of the celestial windows is open!
> The Stairway to Heaven is open;
> The Steps of Light are revealed . . .
> The double Doors to Heaven are open;
> The double doors of *Khebhu* are open
>   for Horus of the east,
>   at daybreak.

Ape-gods symbolizing the waning moon ("Daybreak") begin to pronounce magical "words of power which will cause splendor to issue from the Eye of Horus." The "radiance"—reported earlier as the hallmark of the twin-peaked Mountain of Light—intensifies:

> The sky-god
>> has strengthened the radiance for the king
>> that the king may lift himself to Heaven
>> like the Eye of Ra.
> The king is in this Eye of Horus,
>> where the command of the gods is heard.

The "Eye of Horus" begins to change hues: first it is blue, then it is red. There are excitement and much activity all around:

> The red-Eye of Horus is furious in wrath,
>> its might no one can withstand.
> His messengers hurry, his runner hastens.
> They announce to him who lifts up his arm
>> in the East: "let this one pass."
> Let the god command the fathers, the gods:
> "Be silent . . . lay your hands upon your mouth . . .
>> stand at the doorway of the horizon,
>> open the double doors (of heaven)."

The silence is broken; now there are sound and fury, roaring and quaking:

> The Heaven speaks, the Earth quakes;
> The Earth trembles;
> The two districts of the gods shout;
> The ground is come apart . . .
> When the king ascends to Heaven
>> when he ferries over the vault (to Heaven) . . .

The Earth laughs, the Sky smiles
    when the king ascends to Heaven.
Heaven shouts in joy for him;
The Earth quakes for him.
The roaring tempest drives him,
    it roars like Seth.
The guardians of Heaven's parts
    open the doors of Heaven for him.

Then "the two mountains divide," and there is a lift-off into
a cloudy sky of dawn from which the stars of night are gone:

The sky is overcast,
    the stars are darkened.
The bows are agitated,
    the bones of Earth quake.

Amid the agitation, quaking and thundering, the "Bull of
Heaven" ("whose belly is full of magic") rises from the "Isle
of Flame." Then the agitation ceases; and the king is aloft—
"dawning as a falcon":

They see the king dawning as a falcon,
    as a god;
To live with his fathers,
    to feed with his mothers . . .
The king is a Bull of Heaven . . .
    whose belly is full of magic
    from the Isle of Flame.

Utterance 422 speaks eloquently of this moment:

O this Pepi!
Thou hast departed!
Thou art a Glorious One,
    mighty as a god, seated as Osiris!

Thy soul is within thee;
Thy Power ("Control") has thou behind thee;
The*your head,*
    the Misut-crown is at thy hand . . .
Thou ascendest to thy mother, goddess of Heaven
She lays hold of thine arm,
    she shows thee the way to the horizon,
    to the place where Ra is.
The double doors of heaven are opened for thee,
The double doors of the sky are opened for thee . . .
Thou risest, O Pepi . . . equipped as a god.

(An illustration in the tomb of Ramses IX suggests that the Double Doors were opened by inclining them away from each other; this was achieved by the manipulation of wheels and pulleys, operated by six gods at each door. Through the funnel-like opening, a giant man-like falcon could then emerge. Fig. 25.)

With great self-satisfaction at the achievement, the texts announce to the king's subjects: "He flies who flies; this king Pepi flies away from you, ye mortals. He is not of the Earth, he is of the Heaven . . . This king Pepi flies as a cloud to the sky, like a masthead bird; this king Pepi kisses the sky like a falcon; he reaches the sky of the Horizon god." The king,

Fig. 25

the Pyramid Texts continue, is now "on the Sky-Bearer, the upholder of the stars; from within the shadow of the Walls of God, he crosses the skies."

The king is not simply skyborne, he is orbiting Earth:

> He encompasses the sky like Ra,
> He traverses the sky like Thoth . . .
> He traveleth over the regions of Horus,
> He traveleth over the regions of Seth . . .
> He has completely encircled twice the heavens,
> He has revolved about the two lands . . .
> The king is a falcon surpassing the falcons;
> He is a Great Falcon.

(A verse also states that the king "crosses the sky like *Sunt*, which crosses the sky nine times in one night"; but the meaning of *Sunt* and thus the comparison are as yet undeciphered.)

Still sitting between "these two companions who voyage over the sky," the king soars toward the eastern horizon, far far away in the heavens. His destination is the *Aten*, the Winged Disk, which is also called the Imperishable Star. The prayers now focus on getting the king to the *Aten* and his safe arrival upon it: "*Aten*, let him ascend to thee; enfold him in thine embrace," the texts intone in behalf of the king. There is the abode of Ra, and the prayers seek to assure a favorable welcome for the king, by presenting his arrival at the Celestial Abode as the return of a son to his father:

> Ra of the Aten,
> Thy son has come to thee;
> Pepi comes to thee;
> Let him ascend to thee;
> Enfold him in thy embrace.

Now "there is clamor in Heaven: 'We see a new thing' say the celestial gods; 'a Horus is in the rays of Ra.'" The king—"on his way to Heaven, on the wind"—"advances in

Heaven, he cleaves its firmament," expecting a welcome at his destination.

The celestial journey is to last eight days: "When the hour of the morrow comes, the hour of the eighth day, the king will be summoned by Ra"; the gods who guard the entrance to the *Aten* or to Ra's abode there will let him through, for Ra himself shall await the king on the Imperishable Star:

> When this hour of the morrow comes . . .
> When the king shall stand there, on the star
>   which is on the underside of the Heaven,
>   he shall be judged as a god,
>   listened to like a prince
> The king shall call out to them;
> They shall come to him, those four gods
>   who stand on the *Dam*-scepters of Heaven,
>   that they may speak the king's name to Ra,
>   announce his name Horus of the Horizons:
> "He has come to thee!
> "The king has come to thee!"

Traveling in "the lake that is the heavens," the king nears "the shores of the sky." As he approaches, the gods on the Imperishable Star indeed announce as expected: "The arriver comes . . . Ra has given him his arm on the Stairway to Heaven. 'He Who Knows The Place' comes, say the gods." There, at the gates of the Double Palace, Ra is indeed awaiting the king:

> Thou findest Ra standing there;
> He greets thee, lays hold on thy arm;
> He leads thee into the celestial Double Palace;
> He places thee upon the throne of Osiris.

And the texts announce: "Ra has taken the king to himself, to Heaven, on the eastern side of Heaven . . . the king is on that star which radiates in Heaven."

Now there is one more detail left to accomplish. In the company of "Horus of the *Duat*," described as "the great green divine falcon," the king sets out to find the Tree of Life in the midst of the Place of Offering. "This king Pepi goes to the Field of Life, the birthplace of Ra in the heavens. He finds Kebehet approaching him with these four jars with which she refreshes the heart of the Great God on the day when he awakes. She refreshes the heart of this king Pepi therewith to Life."

Mission achieved, the texts announce with glee:

> Ho, this Pepi!
> All satisfying life is given to thee;
> "Eternity is thine," says Ra . . .
>
> Thou perishest not, thou passest not away
>   for ever and ever.

The king has ascended the Stairway to Heaven; he has reached the Imperishable Star; "his lifetime is eternity, its limit everlastingness."

V

The Gods Who Came
to Planet Earth

Nowadays, we take space flight for granted. We can read of plans for permanently orbiting space settlements without blinking an eye; the development of a reusable space shuttle is viewed not with wonderment, but with appreciation of its cost-saving potentialities. All this, of course, because we have seen with our own eyes, in print and on television, astronauts fly in space and unmanned craft land on other planets. We accept space travel and interplanetary contacts because we have heard with our own ears a mortal named Neil Armstrong, commander of the Apollo 11 spacecraft, report on his radio—for all the world to hear—the first landing by Man on another celestial body, the Moon:

> Houston!
> Tranquility Base here.
> The *Eagle* has landed!

Eagle was not only the code-name for the Lunar Module, but the epithet by which the Apollo 11 spacecraft was called, and the proud nickname by which the three astronauts identified themselves (Fig. 26). The *Falcon* too has soared into space, and landed on the Moon. In the immense National Air and Space Museum of the Smithsonian Institution in Washington, one can see and touch the actual spacecraft that were flown or that were used as backup vehicles in the American

Fig. 26

space program. In a special section where the Moon landings have been simulated with the aid of the original equipment, the visitor can still hear a recorded message from the surface of the Moon:

> O.K., Houston.
> The *Falcon* is on the plain at Hadley!

Whereupon the Manned Spacecraft Center at Houston announced to the world: "That was a jubilant Dave Scott reporting Apollo 15 on the plain at Hadley."

Up to a few decades ago, the notion that a common mortal can put on some special clothes, strap himself in the front part of a long object, then zoom off the face of Earth, seemed preposterous or worse. A century or two ago, such a notion would not have even come about, for there was nothing in human experience or knowledge to trigger such fantasies.

Yet, as we have just described, the Egyptians—5,000 years ago—could readily visualize all this happening to their Pharaoh: he would journey to a launch site east of Egypt; he would enter a subterranean complex of tunnels and chambers; he would safely pass by the installation's atomic plant and radiation chamber. He would don the suit and gear of an astronaut, enter the cabin of an Ascender, and sit strapped between two gods. And then, as the double-doors would open, and the dawn skies would be revealed, the jet engines would ignite and the Ascender would turn into the Celestial Ladder by which the Pharaoh will reach the Abode of the Gods on their "Planet of Millions of Years."

On what TV screens had the Egyptians seen such things happen, that they so firmly believed that all this was really possible?

In the absence of television in their homes, the only alternative would have been to either go to the Spaceport and watch the rocketships come and go, or visit a "Smithsonian" and see the craft on display, accompanied by a knowing guide or viewing flight simulations. The evidence suggests that the ancient Egyptians had indeed done that: they had seen the launch site, and the hardware, and the astronauts with their own eyes. But the astronauts were not Earthlings going elsewhere: they were, rather, astronauts from elsewhere who had come to Planet Earth.

Greatly enamored with art, the ancient Egyptians depicted in their tombs what they had seen and experienced in their lifetimes. The architecturally detailed drawings of the subterranean corridors and chambers of the *Duat* come from the tomb of Seti I. An even more startling depiction has been found in the tomb of Huy, who was viceroy in Nubia and in the Sinai peninsula during the reign of the renowned Pharaoh Tut-Ankh-Amon. Decorated with scenes of people, places and objects from the two domains of which he was viceroy, his tomb preserved to this very day a depiction in vivid colors of a rocketship: its shaft is contained in an underground silo, its upper stage with the command module is above ground (Fig. 27). The shaft is subdivided, like a multi-stage rocket. Inside its lower part, two persons attend to hoses and levers; there is a row of circular dials above them. The silo cutaway shows that it is surrounded by tubular cells for heat-exchange or some other energy-related function.

Above ground, the hemispherical base of the upper stage is clearly depicted in the color painting as scorched, as

Fig. 27

though from a re-entry into Earth's atmosphere. The command module—large enough to hold three to four persons—is conical in shape, and there are vertical "peep holes" all around its bottom. The cabin is surrounded by worshippers, in a landscape of date palm trees and giraffes.

The underground chamber is decorated with leopard skins, and this provides a direct link with certain phases in the Pharaoh's Journey to Immortality. The leopard skin was the distinctive garb symbolically worn by the *Shem* priest as he performed the Opening of the Mouth ceremony. It was the

distinctive garb symbolically worn by the gods who towed
the Pharaoh through "The Secret Path of the Hidden Place"
of the *Duat*—a symbolism repeated to stress the affinity between the Pharaoh's journey and the rocketship in the underground silo.

As the Pyramid Texts make clear, the Pharaoh, in his
Translation into an eternal Afterlife, embarked on a journey
simulating the gods. Ra and Seth, Osiris and Horus and other
gods had ascended to the heavens in this manner. But, the
Egyptians also believed, it was by the same Celestial Boat
that the Great Gods had come down to Earth in the first place.
At the city of An (Heliopolis), Egypt's oldest center of worship, the god Ptah built a special structure—a "Smithsonian
Institution," if you will—wherein an actual space capsule
could be viewed and revered by the people of Egypt!

The secret object—the *Ben-Ben*—was enshrined in the
Het Benben, the "Temple of the Benben." We know from
the hieroglyphic depiction of the place's name that the structure looked like a massive launch tower from within which a
pointed rocket was poised skyward (Fig. 28).

The *Ben-Ben* was, according to the ancient Egyptians, a
solid object that had actually come to Earth from the Celestial Disk. It was the "Celestial Chamber" in which the great
god Ra himself had landed on Earth; the term *Ben* (literally:
"That Which Flowed Out") conveying the combined meanings of "to shine" and "to shoot up in the sky."

An inscription on the stela of the Pharaoh Pi-Ankhi (per
Brugsch, *Dictionnaire Géographique de l'Ancienne Égypte*)
said thus:

> The king Pi-Ankhi mounted the stairs toward the large
> window, in order to view the god Ra within the *Ben-
> Ben*. The king personally, standing up and being all
> alone, pushed apart the bolt and opened the two door-
> leaves. Then he saw his father Ra in the splendid sanc-
> tuary of *Het-Benben*. He saw the *Maad*, Ra's Barge;
> and he saw *Sektet*, the Barge of the *Aten*.

Fig. 28

The shrine, we know from the ancient texts, was guarded and serviced by two groups of gods. There were those "who are outside the *Het-Benben*" but were allowed into the shrine's most sacred parts, for it was their task to receive the offerings from the pilgrims and bring them into the temple. The others were primarily guardians, not only of the *Ben-Ben* itself, but of all "the secret things of Ra which are in *Het-Benben.*" Much as tourists nowadays flock to the Smithsonian to view, admire and even touch the actual vehicles flown in space, so did the devout Egyptians make pilgrimages to Heliopolis, to revere and pray to the *Ben-Ben* —probably with a religious fervor akin to that of the faithful Muslims who make pilgrimages to Mecca, there to pray at the *Qa'aba* (a black stone believed to be a replica of God's "Celestial Chamber").

At the shrine, there was a fountain or well, whose waters acquired a reputation for their healing powers, especially in matters of virility and fertility. The term *Ben* and its hieroglyphic depiction $\bigwedge$ in time indeed acquired the connotations virility and reproduction; and could well have been the source of the meaning "male offspring" that *Ben* has in Hebrew. In addition to virility and reproduction, the shrine also acquired the attributes of rejuvenation; this in turn gave rise to the legend of the *Ben* bird, which the Greeks who had visited Egypt called the *Phoenix*. As these legends had it, the Phoenix was an eagle with plumage partly red and partly golden; once every 500 years, as it was about to die, it went to Heliopolis and in some manner rose again from the ashes of itself (or of its father).

Heliopolis and its healing waters remained venerated until early Christian times; local traditions claim that when Joseph and Mary escaped to Egypt with the child Jesus, they rested by the shrine's well.

The shrine at Heliopolis, Egyptian histories tell, was destroyed several times by enemy invaders. Nothing remains of it nowadays; the *Ben-Ben* is also gone. But it was depicted on Egyptian monuments as a conical chamber within which a god could be seen. Archaeologists have in fact found a stone scale-model of the *Ben-Ben,* showing a god at its open hatch-door in a gesture of welcome (Fig. 29). The true shape of the Celestial Chamber was probably accurately depicted in the tomb of Huy (Fig. 27); that modern command modules—the capsules housing the astronauts atop rocketships at launching, and in which they splash down back to Earth—Fig. 30—look so similar to the *Ben-Ben,* is no doubt a result of similarity of purpose and function.

In the absence of the *Ben-Ben* itself, is there any other physical piece of evidence—and not mere drawings or scale models—left from the Heliopolitan shrine? We have noted above that according to Egyptian texts there were other secret things of Ra on display or in safekeeping at the shrine. In the Book of the Dead nine objects affiliated with the hieroglyph for *Shem* were depicted in the division paralleling the shrine of Heliopolis; it could well be that there were indeed another nine space-related objects or spacecraft parts on display at the shrine.

Archaeologists may also have found a replica of one of these smaller objects. It is an oddly shaped circular object full of intricate curves and cutouts (Fig. 31a); it has baffled all scholars since its discovery in 1936. It is important to realize that the object was found—among other "unusual copper objects"—in the tomb of the crown prince Sabu, son of King Adjib of the First Dynasty. It is, therefore, certain that the object was placed in the tomb circa 3100 B.C. It could have been older, but certainly not more recent, than that date.

Reporting on the discoveries in northern Saqqara (just

Fig. 29   Fig. 30

south of the great pyramids at Gizah), Walter B. Emery *(Great Tombs of the First Dynasty)* described the object as a "bowl-like vessel of schist," and remarked that "No satisfactory explanation of the curious design of this object has been forthcoming." The object was carved from a solid block of schist—a rock which is very brittle and which easily splits into thin, irregular layers. If it were put to any use, it would have quickly broken apart; so the particular stone was chosen because the very unusual and delicate shape could best be carved out in such a material—a means to preserve the shape, rather than to actually use it. This has led other scholars, such as Cyril Aldred *(Egypt to the End of the Old Kingdom)* to conclude that the stone object "possibly imitates a form originally made in metal."

But what metal could have been used in the fourth millenium B.C. to produce the object, what process of precision grinding, what skilled metallurgists were then available to create such a delicate and structurally complex design? And, above all, for what purpose?

A technical study of the object's unique design (Fig. 31b) shed little light on its use or origin. The round object, some twenty-four inches in diameter and less than four inches at its thickest part, was obviously made to fit over a shaft and rotate around an axle. Its three oddly curving cutouts suggest a possible immersion in a liquid during rotation.

There was no further effort made after 1936 to unravel the object's enigma. But its possible function suddenly sprang to our mind in 1976 on reading in a technical magazine of some revolutionary designs of a flywheel being developed in California in connection with the American space program. The flywheel, attached to a rotating shaft of a machine or an engine, has been in use for less than two centuries as a means of regulating the speed of machinery, as well as for accumulating energy for a single spurt, such as in a metal press (or more recently in aviation).

As a rule, flywheels have had heavy rims, for the energy is stored in the wheel's circumference. But in the 1970s, engineers of the Lockheed Missile & Space Company came up with an opposite design—a light-rimmed wheel, claiming it is best suited for saving energy in mass transit trains or storing energy in electrically powered trolley-buses. The research was continued by the Airesearch Manufacturing Company; the model they developed—but never finally perfected—was to be hermetically sealed within a housing filled with lubricant. That their revolutionary flywheel (Fig. 32) looks like the 5,000-year-old object discovered in Egypt is only less amazing than the fact that the perfected object from 3100 B.C. looks like a piece of equipment still in the development stage by space engineers in A.D. 1978!

Where is the metal original of this ancient flywheel? Where are the other objects that were apparently on display at the Heliopolis shrine? Where, for that matter, is the *Ben-Ben* itself? Like so many artifacts whose existence in antiquity has been documented by the ancient peoples beyond doubt, they have all disappeared—destroyed perhaps by natural calamities or wars, perhaps dismantled and taken elsewhere—as war booty, or for safekeeping and hiding away in places long forgotten. Perhaps they were carried back to the heavens; perhaps they are still with us, unrecognized for

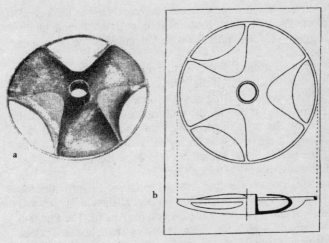

Fig. 31

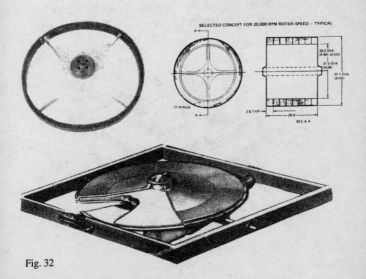

SELECTED CONCEPT FOR 20,000 RPM ROTOR SPEED – TYPICAL

Fig. 32

what they are in some museum basement. Or—as the legend of the Phoenix which connects Heliopolis and Arabia might suggest—hidden under the sealed chamber of the Qa'aba in Mecca . . .

We can surmise, however, that the destruction, disappearance or withdrawal of the shrine's sacred objects had probably taken place during Egypt's so-called First Intermediary Period. In that period, the unification of Egypt came apart and total anarchy reigned. We know that the shrines of Heliopolis were destroyed during the years of disorder; it was then, perhaps, that Ra left his temple at Heliopolis and became *Amon*—the "Hidden God."

When order was first restored in Upper Egypt under the Eleventh Dynasty, the capital was established at Thebes and the supreme god was called Amon (or Amen). The Pharaoh Mentuhotep (Neb-Hepet-Ra) built a vast temple near Thebes, dedicated it to Ra, and topped it with a huge "pyramidion" to commemorate Ra's Celestial Chamber (Fig. 33).

Soon after 2000 B.C., as the Twelfth Dynasty began its reign, Egypt was reunited, order was restored, and access

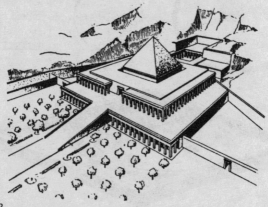

Fig. 33

to Heliopolis was regained. The dynasty's first Pharaoh, Amen-Em-Hat I, immediately undertook the rebuilding of the temples and shrines of Heliopolis; but whether he could also restore the original artifacts enshrined there, or had to do with their stone simulations, no one can say for certain. His son, the Pharaoh Sen-Usert (Kheper-Ka-Ra)—the Sesostris or Sesonchusis of Greek historians—erected in front of the temple two huge granite columns (over sixty-six feet high). On top they were surmounted with a scale replica of Ra's Celestial Chamber—a pyramidion, which was encased in gold or white copper (electrum). One of these granite obelisks still stands where it was raised up some 4,000 years ago; the other was destroyed in the twelfth century A.D.

The Greeks called these pillars *obelisks,* meaning "pointed cutters." The Egyptians called them Beams of the Gods. More of them were set up—always in pairs in front of temple gateways (Fig. 34)—during the eighteenth and nineteenth dynasties (some were in the end carted off to New York, London, Paris, Rome). As stated by the Pharaohs, they raised these obelisks in order to "obtain (from the gods) the gift of Eternal Life," to "obtain Life Everlasting." For the obelisks simulated in stone what earlier Pharaohs had seen (and purportedly reached) in the *Duat,* in the Sacred Mountain: the rocketships of the gods (Fig. 35).

Today's tombstones, engraved with the deceased's name so that he be forever remembered, are scaled-down obelisks—a custom rooted in the days when the gods and their spacecraft were an absolute reality.

The Egyptian word for these Celestial Beings was NTR—a term which in the languages of the ancient Near East meant "One Who Watches." The hieroglyphic sign for *Neter* was ⌐; like all hieroglyphic signs, it must have represented originally an actual, visible object. Suggestions by scholars have ranged from an axe on a long handle to an ensign. Margaret

Fig. 35

Fig. 34

A. Murray *(The Splendor That Was Egypt)* has put forward a more current view. Showing that pottery from the earliest, pre-dynastic period was adorned with drawings of boats carrying a pole with two streamers as a standard (Fig. 36), she concluded that "the pole with the two streamers became the hieroglyphic sign for God."

The interesting point about these earliest drawings is that they showed the boats arriving from a foreign land. When the drawings included people, they showed seated rowers commanded by a tall master, distinguished by the horns protruding from his helmet (Fig. 36)—the mark of being a *Neter*.

Pictorially, then, the Egyptians affirmed from the very beginning that their gods had come to Egypt from elsewhere. This confirmed the tales of how Egypt began—that the god Ptah, having come from the south, and having found Egypt inundated, performed great works of dyking and land reclamation and made the land habitable. There was a place in Egyptian geography which they called *Ta Neter*—"Place/Land of the Gods." It was the narrow straits at the southern end of the Red Sea which is now called Bab-el-Mandeb; it was through that strait that the ships bearing the ensign NTR and carrying the horned gods had come to Egypt.

The Egyptian name for the Red Sea was the Sea of UR. The term *Ta Ur* meant the Foreign Land in the East. Henri Gauthier, who compiled the *Dictionnaire des Noms Géographiques* from all the place names in the hieroglyphic texts, pointed out that the hieroglyph for *Ta Ur* "was a symbol which designated a nautical element . . . The sign means that 'You have to go by boat, to the left side.'" Looking at the map of the ancient lands (page 26), we see that a turn left-

Fig. 36

ward as one came from Egypt and passed through the straits
of Bab-el-Mandeb, would take the sailor along the Arabian
peninsula toward the Persian Gulf.

There are more clues. *Ta Ur* literally meant the Land of
UR, and the name *Ur* was not unknown. It was the birthplace
of Abraham, the Hebrew patriarch. Descended of *Shem,* the
elder son of Noah (the biblical hero of the Deluge), he was
born to his father Terah at the city of Ur, in Chaldea; "and
Terah took Abram his son, and Lot the son of Haran, his
son's son, and Sarah his daughter-in-law, the wife of Abram;
and they went forth from Ur of the Chaldees, to go into the
Land of Canaan."

When archaeologists and linguists began to unravel, at the
beginning of the nineteenth century, the history and written
records of Egypt, Ur was unknown from any other source
except the Old Testament. Chaldea, however, was known:
it was the name by which the Greeks had called Babylonia,
Mesopotamia's ancient kingdom.

The Greek historian Herodotus, who had visited Egypt
and Babylonia in the fifth century B.C., found many similarities in the customs of the Egyptians and the Chaldeans. Describing the sacred precinct of the supreme god *Bel* (whom
he called Jupiter Belus) in the city of Babylon, and its huge
stage tower, he wrote that "on the upmost tower there is a
spacious temple, and inside the temple stands a couch of unusual size, richly adorned, with a golden table by its side.
There is no statue of any kind set up in the place, nor is the
chamber occupied by nights by anyone but a native woman,
who, as the Chaldeans, the priests of this god, affirm, is chosen for himself by the deity. . . . They also declare . . . that
the god comes down in person into this chamber, and sleeps
upon the couch. This is like the story of the Egyptians of
what takes place in their city Thebes, where the woman always passes the night in the temple of the Theban Jupiter
(Amon)."

The more nineteenth-century scholars learnt of Egypt, and
matched the emerging historic picture with the writings of

Greek and Roman historians, the more did two facts stand out: First, that Egyptian civilization and greatness were not like an isolated flower blooming in a cultural desert, but part of overall developments throughout the ancient lands. And secondly, that the biblical tales of other lands and kingdoms, of fortified cities and trade routes, of wars and treaties, of migrations and settlements—were not only true, but also accurate.

The Hittites, known for centuries only from brief mentions in the Bible, were discovered in Egyptian records as mighty adversaries of the Pharaohs. A totally unknown page of history—a pivotal battle between Egyptian armies and Hittite legions which came from Asia Minor, that had taken place at Kadesh in northern Canaan—was discovered not only in text, but also depicted pictorially on temple walls. There was even a historical personal touch, for the Pharaoh ended up marrying the daughter of the Hittite king in an effort to cement peace between them.

Philistines, "People of the Sea," Phoenicians, Hurrians, Amorites—peoples and kingdoms all vouched for, until then, only by the Old Testament—began to emerge as historical realities as the archaeological work progressed in Egypt, and began to spill over into the other lands of the Bible. Greatest of all, however, appeared to have been the veritable ancient empires of Assyria and Babylonia; but where were their magnificent temples, and other remains of their grandeur? And where were their records?

All that travelers had reported from the Land Between The Two Rivers, the vast plain between the Euphrates and the Tigris, were mounds—*tells* in Arabic and Hebrew. In the absence of stone, even the grandest structures of ancient Mesopotamia were built of mud bricks; wars, weather and time reduced them to heaps of soil. Instead of monumental structures, these lands yielded occasional finds of small artifacts; among them there often were tablets of baked clay inscribed with wedge-like markings. Back in 1686, a traveler named Engelbert Kampfer visited Persepolis, the Old Persian capital of the kings whom Alexander had fought. From monuments there, he copied signs and symbols in such a wedge-shaped or cuneiform script, as on the royal seal of Darius (Fig. 37). But he thought that they were only decorations. When it was finally realized that these were inscriptions, no one knew what their language was and how they could be deciphered.

As in the case of the Egyptian hieroglyphs, so it was with the cuneiform writings: The key to the solution was a trilingual inscription. It was found carved on the rocks of forbidding mountains, at a place in Persia called Behistun. In 1835, a major in the British Army, Henry Rawlinson, managed to copy the inscription, and thereafter decipher the script and its languages. As it turned out, the tri-lingual rock inscription was in Old Persian, Elamite and Akkadian. Akkadian was the mother-language of all the Semitic languages; and it was through the knowledge of Hebrew that scholars were able to read and understand the Mesopotamian inscriptions of the Assyrians and the Babylonians.

Spurred by such discoveries, a Paris-born Englishman named Henry Austen Layard reached Mosul, a caravan junction in northeastern Iraq (then part of the Ottoman-Turkish Empire) in 1840. There he was the guest of William F. Ainsworth, whose *Researches in Assyria, Babylonia and Chaldea* (1838)—along with earlier reports and small finds by

Fig. 37

Claudius J. Rich *(Memoir on the Ruins of Babylon)*—not only fired Layard's imagination, but also led to scientific and monetary support from the British Museum and the Royal Geographical Society. Versed both in the pertinent biblical references and the Greek classical writings, Layard kept recalling that an officer in Alexander's army reported seeing in the area "a place of pyramids and remains of an ancient city"—a city whose ruins were considered ancient even in Alexander's times!

His local friends showed him the various *tells* in the area, indicating that there were ancient cities buried beneath them. His excitement grew most when he reached a place called *Birs Nimrud.* "I saw for the first time the great conical mound of Nimrud rising against the clear evening sky," he later wrote in his *Autobiography.* "The impression that it made upon me was one never to be forgotten." Was it not the place of the buried pyramid seen by Alexander's officer? And surely was the place associated with the biblical Nimrod, "the mighty hunter by the grace of the Lord," who launched the kingdoms and royal cities of Mesopotamia (Genesis X)—

> And the beginning of his kingdom: *Babel* and *Erech* and *Akkad,* all in the Land of *Shin'ar;* Out of that land there emanated *Ashur,* where *Nineveh* was built—a city of wide streets; and *Khalah,* and *Ressen.*

With the support of Major Rawlinson, who by then was the British Resident and Consul in Baghdad, Layard returned in 1845 to Mosul to begin digging at his cherished Nimrud. But whatever he was to find—and find he did—the claim to be the first modern archaeologist in Mesopotamia was not to be his. Two years earlier, Paul-Emile Botta, the French Consul at Mosul (whom Layard had met and befriended) proceeded to excavate a mound somewhat north of Mosul, on the other side of the Tigris River. The natives called the place Khorsabad; the cuneiform inscriptions uncovered there identified it as *Dur-Sharru-Kin,* the ancient capital of the biblical Sargon, king of Assyria. Commanding the vast city and its palaces and temples was indeed a pyramid constructed in seven steps, the term for which is ziggurat (Fig. 38).

Spurred by Botta's discoveries, Layard began to dig at his chosen mound, where he believed he would uncover *Nineveh,* the Assyrian capital of biblical renown. Though the place turned out to be the Assyrian military center named Kalhu (the biblical *Khala*), the treasures that were uncovered were worth the effort. They included an obelisk set up by King Shalmaneser II, on which he listed among those paying him tribute "Jehu, son of Omri, king of Israel" (Fig. 39).

Assyrian finds now directly confirmed the historical veracity of the Old Testament.

Encouraged, Layard began to excavate in 1849 a mound directly opposite Mosul, on the eastern banks of the Tigris. The place, locally named Kuyunjik, was indeed Nineveh—the capital established by Sennacherib, the Assyrian king whose army was smitten by the Lord's angel when he besieged Jerusalem (II Kings 18). After him, Nineveh served as the capital of Esarhaddon and Ashurbanipal. The art treasures carted off from there to the British Museum still make up the most impressive portion of its Assyrian displays.

As the pace of excavations gathered momentum, as archaeological teams from other nations joined the race, all the Assyrian and Babylonian cities named in the Bible (with one minor exception) were uncovered. But as the world's museums filled up with the ancient treasures, the most important finds were the simple clay tablets—some small enough to be held in the palm of the scribe's hand—on which the Assyrians, Babylonians and other peoples of western Asia wrote down commercial contracts, court rulings, marriage and inheritance records, geographical lists, mathematical information, medical formulas, laws and regulations, royal histories—indeed, every aspect of life by advanced and highly civilized societies. Epic tales, Creation tales, proverbs, philosophical writings, love songs

Fig. 38

Fig. 39

and the like made up a vast literary heritage. And there were matters celestial—lists of stars and constellations, planetary information, astronomical tables; and lists of gods, their family relationships, their attributes, their tasks and functions—gods headed by twelve Great Gods, "Gods of Heaven and Earth," with whom there were associated the twelve months, the twelve constellations of the Zodiac, and twelve members of our solar system.

As the inscriptions themselves occasionally stated, their language stemmed from the Akkadian. This and other evidence confirmed the biblical narrative, that Assyria and Babylon (which appeared on the historical stage circa 1900 B.C.) were preceded by a kingdom named Akkad. It was founded by *Sharru-Kin*—"the Righteous Ruler"—whom we call Sargon I, circa 2400 B.C. Some of his inscriptions were also found; in them he boasted that by the grace of his god *Enlil,* his empire stretched from the Persian Gulf to the Mediterranean Sea. He called himself "King of *Akkad,* King of *Kish*"; and he claimed to have "defeated *Uruk,* tore down its wall . . . (was) victorious in battle with the inhabitants of *Ur.*"

Many scholars believe that Sargon I was the biblical Nimrod, so that the biblical verses apply to him and to a capital named Kish (or Kush by biblical spelling) where kingship existed even before Akkad:

> And Kush begot Nimrod;
> He was first to be a Mighty Man in the land . . .
> And the beginning of his kingdom:
> Babel and Erech and Akkad,
> All in the Land of Shin'ar.

The royal city of Akkad was discovered southeast of Babylon; the ancient city of Kish was also discovered, southeast of Akkad. Indeed, the more archaeologists moved down in the plain between the two rivers, in a southeasterly direction, the greater was the antiquity of the places unearthed.

At a place now called Warka, the city of Uruk, which Sargon I claimed to have defeated, the biblical *Erech*, was found. It took the archaeologists from the third millenium B.C. to the *fourth* millenium B.C.! There, they found the first-ever pottery baked in a kiln; evidence of the first-ever use of a potter's wheel; a pavement of limestone blocks which is the oldest of its kind; the first-ever ziggurat (step pyramid); and the world's first written records: inscribed texts (Fig. 40) and engraved cylinder seals (Fig. 41) which, when rolled on wet clay, left a permanent imprint.

Ur—birthplace of Abraham—was also found, farther south, where the coastline of the Persian Gulf had reached in antiquity. It was a great commercial center, site of a huge ziggurat, the seat of many dynasties. Was then the southern, more ancient part of Mesopotamia, the biblical Land of *Shin'ar*—the place where the events of the Tower of Babel had taken place?

One of the greatest discoveries in Mesopotamia was the library of Ashurbanipal in Nineveh, which contained more than 25,000 tablets arranged by subject. A king of great culture, Ashurbanipal collected every text he could lay his hands on, and in addition set his scribes to copy and translate texts otherwise unavailable. Many tablets were identified by the scribes as "copies of olden texts." A group of twenty-three tablets, for example, ended with the postscript: "twenty-third tablet; language of *Shumer* not changed." Ashurbanipal himself stated in an inscription:

> The god of scribes has bestowed on me the gift of the knowledge of his art. I have been initiated into the secrets of writing. I can even read the intricate tablets in *Shumerian*. I understand the enigmatic words in the stone carvings from the days before the Deluge.

In 1853, Henry Rawlinson suggested to the Royal Asiatic Society that there possibly was an unknown language that preceded Akkadian, pointing out that the Assyrian and

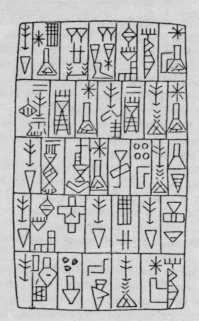

Fig. 40

Fig. 41

Babylonian texts often used words borrowed from that unknown language, especially in scientific or religious texts. In 1869 Jules Oppert proposed at a meeting of the French Society of Numismatics and Archaeology that recognition be given to the existence of such an early language and of the people who spoke and wrote it. He showed that the Akkadians called their predecessors *Shumerians,* and spoke of the Land of *Shumer* (Fig. 42).

It was, in fact, the biblical Land of *Shin'ar.* It was the land whose name—*Shumer*—literally meant Land of the Watchers. It was indeed the Egyptian *Ta Neter*—Land of the Watchers, the land from which the gods had come to Egypt.

As difficult as it was at the time, scholars have accepted, after the grandeur and antiquity of Egypt had been unearthed, that civilization (as known to the West) did not begin in Rome and Greece. Could it now be, as the Egyptians themselves had suggested, that civilization and religion began not in Egypt, but in southern Mesopotamia?

In the century that followed the first Mesopotamian discoveries, it has become evident beyond doubt that it was indeed in Sumer (scholars find this spelling easier to pronounce) that modern Civilization (with a capital 'C') began. It was there, soon after 4000 B.C.—nearly 6,000 years ago—that all the essential elements of a high civilization suddenly blossomed out, as though from nowhere and for no apparent reason. There is hardly any aspect of our present culture and civilization whose roots and precursors cannot be found in Sumer: cities, high-rise buildings, streets, marketplaces, granaries, wharves, schools, temples; metallurgy, medicine, surgery, textile making, gourmet foods, agriculture, irrigation; the use of bricks, the invention of the kiln; the first-ever wheel, carts; ships and navigation; international trade; weights and measures; kingship, laws, courts, juries; writing and recordkeeping; music, musical notes, musical instru-

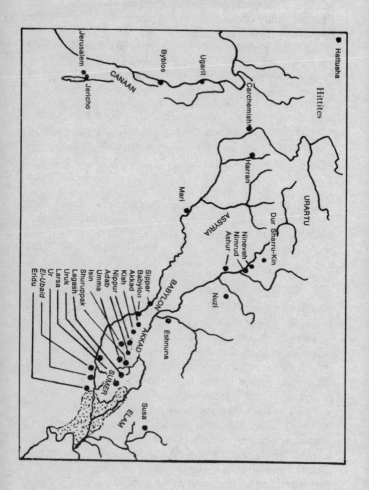

Fig. 42

ments, dance and acrobatics; domestic animals and zoos; warfare, artisanship, prostitution. And above all: the knowledge and study of the heavens, and the gods "who from the Heavens to Earth had come."

Let it be clarified here that neither the Akkadians nor the Sumerians had called these visitors to Earth gods. It is through later paganism that the notion of divine beings or gods has filtered into our language and thinking. When we employ the term here, it is only because of its general acceptance and usage that we do so.

The Akkadians called them *Ilu*—"Lofty Ones"—from which the Hebrew, biblical *El* stems. The Canaanites and Phoenicians called them *Ba'al*—Lord. But at the very beginning of all these religions, the Sumerians called them DIN. GIR, "the Righteous Ones of the Rocketships." In the early pictographic writing of the Sumerians (which was later stylized into cuneiform wedge-writing), the terms DIN and GIR were written ⊏⊐ ⋙ When the two are combined we can see that the cutter or GIR—shaped like a conical-pyramidical command module—fits perfectly into the nose of the DIN, pictured as a multi-stage rocket. Moreover, if we stand the completed word-picture up, we find that it is amazingly similar to the rocket ship in the underground silo depicted in the Egyptian tomb of Huy (Fig. 43).

From the Sumerian cosmological tales and epic poems, from texts that served as autobiographies of these gods, from lists of their functions and relationships and cities, from chronologies and histories called King Lists, and a wealth of other texts, inscriptions and drawings, we have pieced together a cohesive drama of what had happened in prehistoric times, and how it all began.

Their story begins in primeval times, when our solar system was still young. It was then that a large planet appeared from outer space and was drawn into the solar system. The Sumerians called the invader NIBIRU—"Planet of the Crossing"; the Babylonian name for it was *Marduk*. As it

Fig. 43

passed by the outer planets, Marduk's course curved in, to a collision course with an old member of the solar system—a planet named Tiamat. As the two came together, the satellites of Marduk split Tiamat in half. Its lower part was smashed into bits and pieces, creating the comets and the asteroid belt—the "celestial bracelet" of planetary debris that orbits between Jupiter and Mars. Tiamat's upper part, together with its chief satellite, were thrown into a new orbit, to become Earth and the Moon.

Marduk itself, intact, was caught in a vast elliptical orbit around the Sun, returning to the site of the "celestial battle" between Jupiter and Mars once in 3,600 Earth-years (Fig. 44). It was thus that the solar system ended up with *twelve* members—the Sun, the Moon (which the Sumerians considered a celestial body in its own right), the nine planets we know of, and one more—the twelfth: Marduk.

When Marduk invaded our solar system, it brought with it

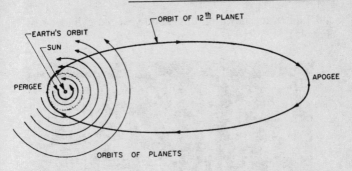

Fig. 44

the seed of life. In the collision with Tiamat, some of the seed of life was transferred to its surviving part—Planet Earth. As life evolved on Earth, it emulated evolution on Marduk. And so it was that when on Earth the human species just began to stir, on Marduk intelligent beings had already achieved high levels of civilization and technology.

It was from that twelfth member of the solar system, the Sumerians said, that astronauts had come to Earth—the "Gods of Heaven and Earth." It was from such Sumerian beliefs, that all the other ancient peoples acquired their religions and gods. These gods, the Sumerians said, created Mankind and eventually gave it civilization—all knowledge, all sciences, including an incredible level of a sophisticated astronomy.

This knowledge encompassed recognition of the Sun as the central body of the solar system, cognizance of all the planets we know of today—even the outer planets Uranus, Neptune and Pluto, which are relatively recent discoveries of modern astronomy—planets which could not have been observed and identified with the naked eye. And, in planetary texts and lists, as well as in pictorial depictions, the Sumerians insisted that there was one more planet—NIBIRU, *Marduk*—which, when nearest Earth, passed between Mars

and Jupiter, as shown on this 4,500-year-old cylinder seal
(Fig. 45).

Fig. 45

The sophistication in celestial knowledge—attributed
by the Sumerians to the astronauts who had come from
Marduk—was not limited to familiarity with the solar system. There was the endless universe, full of stars. It was first-
ever in Sumer—not centuries later in Greece, as has been
thought—that the stars were identified, grouped together
into constellations, given names, and located in the heavens. All the constellations we now recognize in the northern
skies, and most of the constellations of the southern skies,
are listed in Sumerian astronomical tablets—in their correct
order and by names which we have been using to this very
day!

Of the greatest importance were the constellations which
appear to ring the plane or band in which the planets orbit the
Sun. Called by the Sumerians UL.HE ("The Shiny Herd")—
which the Greeks adopted as the *zodiakos kyklos* ("Animal
Circle") and we still call the Zodiac—they were arranged in
twelve groups, to form the twelve Houses of the Zodiac. Not
only the names by which these star groups were called by

the Sumerians—Bull *(Taurus)*, Twins *(Gemini)*, The Pincer *(Cancer)*, Lion *(Leo)* and so exactly on—but even their pictorial depictions have remained unchanged through the millenia (Fig. 46). The much later Egyptian Zodiac representations were almost identical to the Sumerian ones (Fig. 47).

In addition to the concepts of a spherical astronomy that we employ to this very day (including the notions of a celestial axis, poles, ecliptic, equinoxes and the like) which were already perfected in Sumerian times, there was also the astounding familiarity with the phenomenon or Precession. As we now know, there is an illusion of a retardation in Earth's orbit as an observer from Earth pinpoints the Sun on a fixed date (such as the first day of spring) against the Zodiac constellations that act as a backdrop in space. Caused by the fact that the Earth's axis is inclined relatively to its plane of orbit around the Sun, this retardation or Precession is infinitesimal in terms of human lifespans: in seventy-two years, the shift in the Zodiac backdrop is a mere 1° of the 360° Celestial Circle.

Since the Zodiac circle surrounding the band in which Earth (and other planets) orbits around the Sun was divided

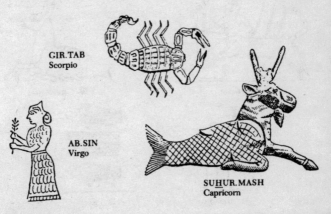

GIR.TAB
Scorpio

AB.SIN
Virgo

SUHUR.MASH
Capricorn

Fig. 46

1. Aries.   2. Taurus.   3. Gemini.

4. Cancer.   5. Leo

6. Virgo.   7. Libra.   8. Scorpio.

9. Sagittarius.   10. Capricorn.

11. Aquarius.   12. Pisces.

Fig. 47

into an arbitrary twelve Houses, each takes up one-twelfth of the full circle, or a celestial space of 30°. It thus takes Earth 2,160 years (72 x 30) to retard through the full span of a Zodiac House. In other words, if an astronomer on Earth has been observing (as is now done) the spring day when the Sun began to rise against the constellation or House of Pisces, his descendants 2,160 years later would observe the event with the Sun against the backdrop of the adjacent constellation, the "House" of *Aquarius.*

No single man, perhaps even no single nation, could have possibly observed, noted and understood the phenomenon in antiquity. Yet the evidence is irrefutable: The Sumerians, who began their time-counting or calendar in the Age of Taurus (which began circa 4400 B.C.), were aware of and recorded in their astronomical lists the previous precessional shifts to Gemini (circa 6500 B.C.), Cancer (circa 8700 B.C.) and Leo (circa 10,900 B.C.)! Needless to say, it was duly recognized circa 2200 B.C. that the first day of spring—New Year to the peoples of Mesopotamia—retarded a full 30° and shifted to the constellation or "Age" of *Aries,* the Ram (KU.MAL in Sumerian).

It has been recognized by some of the earlier scholars who combined their knowledge of Egyptology/Assyriology with astronomy, that the textual and pictorial depictions employed the Zodiac Age as a grand celestial calendar, whereby events on Earth were related to the grander scale of the heavens. The knowledge has been employed in more recent times as a means of prehistoric and historic chronological aid in such studies as that by G. de Santillana and H. von Dechend *(Hamlet's Mill).* There is no doubt, for example, that the Lion-like Sphinx south of Heliopolis, or the Ram-like Sphinxes guarding the temples of Karnak, depicted the Zodiac ages in which the events they stood for had occurred, or in which the gods or kings represented had been supreme.

Central to this knowledge of astronomy, and in consequence to all the religions, beliefs, events and depictions of the ancient world, was thus the conviction that there is one more planet in our solar system, a planet with the vastest orbit, a supreme planet or "Celestial Lord"—the one the Egyptians called the Imperishable Star, or the "Planet of Millions of Years"—the celestial abode of the gods. The ancient peoples, without exception, paid homage to this planet, the one with the vastest, most majestic orbit. In Egypt, in Mesopotamia and elsewhere, its ubiquitous emblem was that of the Winged Globe (Fig. 48).

Recognizing that the Celestial Disk, in Egyptian depictions, stood for the Celestial Abode of Ra, scholars have persisted in referring to Ra as a "Sun God" and to the Winged Disk as a "Sun Disk." It should now be clear that it was not the Sun, but the Twelfth Planet which was so depicted. Indeed, Egyptian depictions clearly distinguished between the Celestial Disk representing this planet, and the Sun. As can be seen (Fig. 49), *both* were shown in the heavens (represented by the arched form of the goddess Nut); clearly, then, two celestial bodies and not a single one are involved. Clearly too, the Twelfth Planet is shown as a celestial globe or disk—a planet; whereas the Sun is shown emitting its benevolent rays (in this instance, on the goddess Hat-Hor, "Lady of the Mines" of the Sinai peninsula).

Did the Egyptians then, as the Sumerians, know thousands of years ago that the Sun was the center of the solar system, and that the system consisted of twelve members? We know that it was so from actual celestial maps depicted on mummy coffins.

A well-preserved one, discovered by H. K. Brugsch in 1857 in a tomb at Thebes (Fig. 50), shows the goddess Nut ("The Heavens") in the central panel (painted atop the coffin), surrounded by the twelve constellations of the Zodiac. On the sides of the coffin, the bottom rows depict the twelve hours of the night and of the day. Then the planets—the Celestial Gods—are shown traveling in their prescribed orbits, the Celestial Barques (the Sumerians called these orbits the "destinies" of the planets).

In the central position, we see the globe of the Sun, emit-

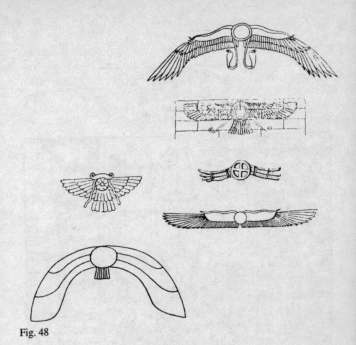

Fig. 48

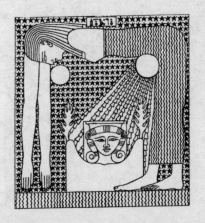

Fig. 49

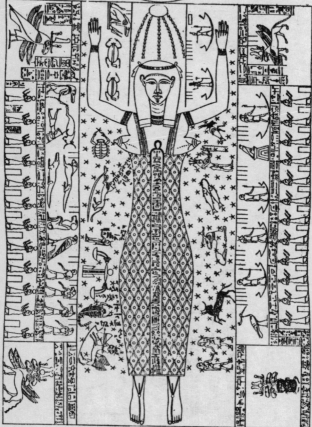

Fig. 50

ting rays. Near the Sun, by Nut's raised left hand, we see two planets: Mercury and Venus. (Venus is correctly depicted as a female—the only planet considered female by all ancient peoples.) Then, in the left-hand panel, we see Earth (accompanied by the emblem of Horus), the Moon, Mars and Jupiter as Celestial Gods traveling in their Celestial Barques.

We see four more Celestial Gods beyond Jupiter, on the right-hand panel. With orbits unknown to the Egyptians (and thus without Barques), we see Saturn, Uranus, Neptune and Pluto. The time of mummification is marked by the Spearman pointing his spear into the midst of the Bull (Taurus).

We thus encounter all the planets in their correct order, including the outer planets whom modern astronomers discovered only rather recently (Brugsch, as others of his time, was unaware of the existence of Pluto).

Scholars who have studied the planetary knowledge in antiquity assumed that the ancient peoples believed that five planets—the Sun being one of them—circled Earth. Any depiction or listing of more planets, these scholars held, was due to some "confusion." But there was no confusion; rather—impressive accuracy: that the Sun was in the system's center, that Earth was a planet, and that in addition to Earth and Moon and the other eight planets known to us today, there is one more large planet. It is depicted atop all others, above the head of Nut, as a major Celestial Lord with its own huge celestial orbit ("Celestial Barque").

Four hundred fifty thousand years ago—according to our Sumerian sources—astronauts from this Celestial Lord landed on Planet Earth.

VI

In the Days before the Deluge

> I understand the enigmatic words in the stone carvings from the days before the Deluge.

So had stated, in a self-laudatory inscription, the Assyrian king Ashurbanipal. Indeed, throughout the diversified literature of ancient Mesopotamia, there were scattered references to a deluge that had swept over Earth. Could it then be, scholars wondered as they came upon such references, that the detailed biblical tale of the Deluge was not a myth or allegory, but the record of an actual event—an event remembered not by the Hebrews alone?

Moreover, even the single sentence in Ashurbanipal's inscription was full of scientific dynamite. He not only confirmed that there had been a Deluge; he stated that his tutoring by the God of Scribes included the understanding of pre-Diluvial inscriptions, "the enigmatic words in the stone carvings from the days before the Deluge." It could only mean that even before the Deluge there had been scribes and stone carvers, languages and writing—that there had been a civilization in the remote days before the Deluge!

It was traumatic enough to have realized that the roots of our modern western civilization go back not to Greece and Judea of the first millenium B.C., and not to Assyria and Babylonia of the second millenium B.C., and not even Egypt of the third millenium B.C.—but to Sumer of the fourth millenium B.C. Now, scientific credibility had to be stretched even farther back, to what even the Sumerians considered "the olden days"—to an enigmatic era "before the Deluge."

Yet, all these shocking revelations should have been old news to anyone who had cared to read the Old Testament's words for what they actually said: that after Earth and the Asteroid Belt (the *Raki'a* or Heaven of Genesis) had been created, and Earth had taken shape, and life evolved, and "the Adam" created—Man was placed in the Orchard that was in Eden. But through the machinations of a brilliant "Serpent" who dared call the bluff of God, Adam and his female companion Eve acquired certain knowing which they were not supposed to possess. Thereupon, the Lord—speaking to unnamed colleagues—grew concerned that Man, "having become as one of us," might also help himself to the Tree of Life, "and eat, and live forever."

> So He drove out the Adam;
> And he placed at the east of the Garden of Eden
> the Cherubim, and the Flaming Sword which revolveth,
> to guard the way to the Tree of Life.

Thus was Adam expelled from the wonderful orchard which the Lord had planted in Eden, from then on to "eat the herbs of the field" and obtain his sustenance "by the sweat of thy face." And "Adam knew Eve his wife and she conceived and bore Cain . . . and she bore again, his brother, Abel; and Abel was a keeper of sheep, and Cain was a tiller of the land."

The biblical claim of a pre-Diluvial civilization then proceeds along two lines, beginning with the line of Cain. Having murdered Abel—there is a hint of homosexuality as the cause—Cain was expelled farther east, to the "Land of Migrations." There his wife bore him Enoch—a name meaning "Foundation"; and the Bible explains that Cain "was building a city" when Enoch was born, "so he named the city 'Enoch,' as a namesake of his son Enoch." (The application of the same name to a person and the city associated with him was a custom that prevailed throughout the history of the ancient Near East.)

The line of Cain continued through Irad, Mechuyah-el, Metusha-el and Lamech. The first son of Lamech was Jabal—a name which in the original Hebrew *(Yuval)* means "The Lute Player." As the Book of Genesis explains, "Jabal was the ancestor of all such as play the harp and the lyre." A second son of Cain, Tubal-Cain, could "sharpen all cutters of copper and iron." What became of these capable people in the eastern Lands of Migration, we are no longer told; for the Old Testament, considering the line of Cain to be accursed, lost all interest in tracking further their genealogies and fate.

Instead, the Book of Genesis (in Chapter V) turns back to Adam and to his third son Seth. Adam, we are told, was 130 years old when Seth was born, and lived another 800 years for a total of 930 years. Seth, who fathered Enosh at age 105, lived to be 912. Enosh begot Cainan at age 90, and died at age 905. Cainan lived to the ripe age of 910; his son Mahalal-el was 895 years old when he died; and his son, Jared, passed on at age 962.

For all these pre-Diluvial patriarchs, the Book of Genesis provides the bare biographical information: who was their father, when their male heir was born, and (after "giving birth to other sons and daughters") when they died. But when the next patriarch is listed, he gets special treatment:

> And Jared lived a hundred and sixty-two years, and begot *Enoch* . . .
> And Enoch lived sixty-five years, and begot Methuselah.
> And Enoch walked with the Lord, after he had begotten Methusaleh, for three hundred years; and he begot (other) sons and daughters.
> And all the days of Enoch were three hundred and sixty and five years.

And here comes the explanation—an astounding explanation—of why Enoch was singled out for so much attention and biographical detail: Enoch did not die!

> For Enoch had walked with the Lord, and was gone;
> for the Lord had taken him away.

Methusaleh lived the longest—969 years—and was succeeded by Lamech. Lamech (who lived to be 777 years) begot Noah—the hero of the Deluge. Here too there is a brief biographical-historical note: Lamech had so named his son, we are informed, because Mankind was undergoing at that time great sufferings, and the earth was barren and unproductive. In naming his son Noah ("Respite"), Lamech expressed the hope that "This one shall bring us respite of our toil and frustrations of the land which the Lord had cursed."

And so, through ten generations of pre-Diluvial patriarchs blessed with what scholars call "legendary" life spans, the biblical narrative reaches the momentous events of the Deluge.

The Deluge is presented in the Book of Genesis as the opportunity seized by the Lord "to destroy the Man whom I had created from the face of the Earth." The ancient authors found it necessary to provide an explanation for such a far-reaching decision. It had to do, we are told, with Man's sexual perversions; specifically, with the sexual relations between "the daughters of Man" and "the sons of the gods."

In spite of the monotheistic endeavors of the compilers and editors of the Book of Genesis, struggling to proclaim faith in a sole deity in a world that in those days believed in many gods, there remain numerous slip-ups where the biblical narrative speaks of gods in the plural. The very term for "deity," (when the Lord is not specifically named as Yahweh), is not the singular *El* but the plural *Elohim.* When the idea of creating Adam occurs, the narrative adopts the plural language: "And Elohim (= the deities) said: 'Let *us* make Man in *our* image and after *our* Likeness.'" And when the incident with the Fruit of Knowing had occurred, Elohim again spoke in the plural to unnamed colleagues.

Now, it transpires from four enigmatic verses of Genesis VI that set the stage for the Deluge, that not only were there deities in the plural, but that they even had sons (in the plural). These sons upset the Lord by having sex with the daughters of Man, compounding their sins by having children or demi-gods born from this illicit lovemaking:

> And it came to pass
> When the Earthlings began to multiply upon the Earth, and daughters were born unto them—
> That the sons of the gods saw the daughters of Adam, that they were good;
> And they took them for wives, of all which they chose.

And the Old Testament explains further:

> The *Nefilim* were upon the Earth in those days and thereafter too;
> Those sons of the gods, who cohabited with the daughters of the Adam, and they bore children unto them.
> They were the Mighty Ones of Eternity, the People of the *Shem*.

Nefilim—traditionally translated "giants"—literally means "Those Who Were Cast Upon" the earth. They were the "Sons of the Gods"—"the people of the *Shem*," the *People of the Rocketships.*

We are back to Sumer and the DIN.GIR, the "Righteous Ones of the Rocket Ships."

Let us then pick up the Sumerian record where we left off—450,000 years ago.

*   *   *

It was some 450,000 years ago, the Sumerian texts claim, that astronauts from Marduk came to Earth in search of gold. Not for jewelry, but for some pressing need affecting survival on the Twelfth Planet.

The first landing party numbered fifty astronauts; they were called *Anunnaki*—"Those of Heaven Who Are on Earth." They splashed down in the Arabian Sea, and made their way to the head of the Persian Gulf, establishing there their first Earth Station E.RI.DU—"Home in Faraway Built." Their commander was a brilliant scientist and engineer who loved to sail the seas, whose hobby was fishing. He was called E.A—"He Whose House Is Water," and was depicted as the prototype Aquarius; but having led the landing on Earth, he was given the title EN.KI—"Lord Earth." Like all the Sumerian gods, his distinguishing feature was the horned headdress (Fig. 51).

The original plan, it appears, was to extract the gold from the seawaters; but the plan proved unsatisfactory. The only alternative was to do it the hard way: to mine ores in southeastern Africa, haul them by ship to Mesopotamia, and there smelt and refine them. The refined gold ingots were then sent aloft in Shuttlecraft, to an Earth-orbiting craft. There they awaited the periodic arrival of the Mother Spaceship, which took the precious metal back home.

To make this possible, more Anunnaki were landed on Earth, until their number reached 600; another 300 serviced the Shuttlecraft and orbiting station. A Spaceport was built at *Sippar* ("Bird Town") in Mesopotamia, on a site aligned with the Near East's most conspicuous landmark—the peaks of Ararat. Other settlements with various functions—such as the smelting and refining center of *Bad-Tibira,* a medical center named *Shuruppak*—were laid out to form an arrow-like Landing Corridor. In the exact center, NIBRU.KI—"The Crossing Place on Earth" (*Nippur* in Akkadian)—was established as the Mission Control Center.

The commander of this expanded enterprise on Planet Earth was EN.LIL—"Lord of the Command." In the early

Sumerian pictographic writing, Enlil's name and his Mission Control Center were depicted as a complex of structures with tall antennas and wide radar screens (Fig. 52).

Both Ea/Enki and Enlil were sons of the then ruler on the Twelfth Planet, AN (Akkadian *Anu*), whose name meant "He of the Heavens" and was written pictographically as a star ⋇ Ea was the firstborn; but because Enlil was born to Anu by another wife who was also his half-sister, Enlil and not Ea was the heir to the throne. Now Enlil was sent to Earth, and took over the command from Ea, the so-called Lord Earth. Matters were further complicated by the despatch to Earth as Chief Medical Officer of NIN.HUR. SAG ("Lady of the Mountainpeak"), a half-sister of both Ea and Enlil—enticing both brothers to seek her favors; for by the same rules of succession, a son to one of them by her

Fig. 51

EN

LIL

Fig. 52

would inherit the throne. The lingering resentment on Ea's part, compounded by the growing competition between the brothers, eventually spilled off to their offspring and was the underlying cause of many events that followed.

As the millenia passed on Earth—though to the Anunnaki each 3,600 years were but one year of their own life cycles—these rank-and-file astronauts began to grumble and complain. Was it indeed their task as spacemen to dig for ores deep inside dark, dusty, hot mines? Ea—perhaps to avoid friction with his brother—spent more and more time in southeastern Africa, away from Mesopotamia. The Anunnaki who toiled in the mines addressed their complaints to him; together they talked over their mutual dissatisfactions.

Then, one day, as Enlil arrived in the mining area for a tour of inspection, the signal was given. A mutiny was declared. The Anunnaki left the mines, put their tools on fire, marched on Enlil's residence and encircled it, shouting: "No more!"

Enlil contacted Anu and offered to resign his command and return to the home planet. Anu came down to Earth. A court-martial was held. Enlil demanded that the instigator of the mutiny be put to death. The Anunnaki, all as one, refused to divulge his identity. Hearing the evidence, Anu concluded that the work was indeed too harsh. Was the gold mining then to be discontinued?

Ea then offered a solution. In southeastern Africa, he said, there roamed a being that could be trained to perform some of the mining tasks—if only the "mark of the Anunnaki" could be implanted upon it. He was talking of the Apemen and Apewomen, who had evolved on Earth—but were still far behind the evolutionary level attained by the inhabitants of the Twelfth Planet. After much deliberation, Ea was given the go-ahead: "Create a *Lulu*," a "primitive worker," he was told; "let him bear the yoke of the Anunnaki."

Ninhursag, the chief medical officer, was to assist him. There was much trial and error until the right procedure was perfected. Extracting the egg of Apewoman, Ea and Ninhursag fertilized it with the sperm of a young astronaut. Then they reimplanted the fertilized egg not in the womb of the Apewoman, but in the womb of a female astronaut. Finally, the "Perfect Model" was achieved, and Ninhursag shouted with joy: "I have created—my hands have made it!" She held up for all to see the first *Homo sapiens* (Fig. 53)—the Earth's first-ever test-tube baby.

But like any hybrid, the Earthling could not procreate on his own. To obtain more primitive workers, Apewomen eggs were extracted, fertilized, and reimplanted in the wombs of "birth goddesses"—fourteen at a time: seven to be born males, seven females. As the Earthlings began to take over mining work in southeastern Africa, the Anunnaki who toiled in Mesopotamia grew jealous: they too clamored for primitive workers. Over the objections of Ea, forcibly, Enlil seized some of the Earthlings and brought them to the E.DIN—the "Abode Of The Righteous Ones" in Mesopotamia. The event is recalled in the Bible: "And the Lord took the Adam, and He placed him in the garden in *Eden,* to till it and tend it."

All along, the astronauts who had come to Earth were preoccupied with the problem of longevity. Their biological clocks were set for their own planet: the time it took their planet to orbit the Sun once, was to them but a single year of their life

Fig. 53

Fig. 54

spans. But in such a single year, Earth orbited the Sun 3,600 times—a span of 3,600 years for Earth-originated life. To maintain their longer cycles on the quick-paced Earth, the astronauts consumed a "Food of Life" and "Water of Life" which were provided from the home planet. At his biological laboratories in Eridu, whose emblem was the sign of the Entwined Serpents (Fig. 54), Ea tried to unravel the secrets of life, reproduction, death. Why did the children born to the astronauts on Earth age so much faster than their parents? Why did Apemen live such short lives? Why did the hybrid *Homo sapiens* live much longer than Apeman, but only brief lives compared to the visitors to Earth? Was it environment, or inherent genetic traits?

Conducting further experiments in genetic manipulation on the hybrids, and using his own sperm, Ea came up with a new "perfect model" of Earthling. *Adapa,* as Ea had named him, had greater intelligence; he acquired the all-important ability to procreate; but not the longevity of the astronauts:

> With wide understanding
>   he had perfected him . . .
> To him he had given Knowing;
> Lasting Life him he did not give.

Thus were Adam and Eve of the Book of Genesis given the gift or fruit not only of Knowledge but also of *Knowing*—the biblical Hebrew term for intercourse for the purpose of having offspring. We find this "biblical" tale illustrated in an archaic Sumerian drawing (Fig. 55).

Enlil was outraged on discovering what Ea had done. It was never intended that Man should be able to procreate like the gods. What next, he asked—would Ea also achieve for Man an everlasting life span? On the home planet, Anu too was perturbed. "Rising from his throne, he ordered: 'Let them fetch Adapa hither!'"

Afraid that his perfected human would be destroyed at the Celestial Abode, Ea instructed him to avoid the food and water that would be offered to him, for they would contain poison; "He gave him this advice:

> Adapa,
> Thou art going before Anu, the Ruler.
> The road to Heaven wilt thou take.
> When to Heaven thou hast gone up,
>   and hast approached the gate of Anu,
> Tammuz and Gizzida at the gate will be standing . . .
> They will speak to Anu;
> Anu's benign face they will cause to be shown thee.

Fig. 55

As thou standest before Anu,
When they offer thee the Bread of Death,
   thou shalt not eat it.
When they offer thee the Water of Death,
   thou shalt not drink it . . .

"Then he made him take the road to Heaven, and to Heaven he went up." When Anu saw Adapa, he was impressed by his intelligence and the extent to which he had learned from Ea "the plan of Heaven and Earth." "What shall we do about him," he asked his counselors, now that Ea "distinguished him by making a *Shem* for him"—by letting Adapa travel in a spacecraft from Earth to Marduk?

The decision was to keep Adapa permanently on Marduk. So that he could survive, "the Bread of Life they brought him," and the Water of Life too. But forewarned by Ea, Adapa refused to eat or drink. When his erroneous reasons were discovered, it was too late; his chance to obtain everlasting life was missed.

Adapa was returned to Earth—a trip during which Adapa saw the "awesomeness" of space, "from the horizon of Heaven to the zenith of Heaven." He was ordained as the High Priest of Eridu; he was promised by Anu that henceforth the Goddess of Healing would also attend to the ailments of Mankind. But Mortal's ultimate goal—everlasting life—was no longer his.

From then on, Mankind proliferated. The humans were no longer just slaves in the mines or serfs in the fields. They performed all tasks, built "houses" for the gods—we call them "temples"—and quickly learned how to cook, dance and play music for them. It was not long before the young Anunnaki, short of female company of their own, took to having sex with the daughters of Man. Since they were all of the same first Seed of Life, and Man was a hybrid created with the genetic "essence" of the Anunnaki, the male astronauts and the female Earthlings discovered that they were biologically compatible; "and children were born unto them."

Enlil viewed these developments with rising apprehension. The original purpose of coming to Earth, the sense of mission, the dedication to the task—were dissipated and gone. The good life seemed to be the main concern of the Anunnaki—and with a race of hybrids to boot!

Nature, as it were, offered Enlil a chance to put a halt to the deteriorating mores and ethics of the Anunnaki. Earth was entering a new ice age, and the pleasant climate was changing. As it got colder, it also became dryer. Rains became less frequent, the river waters sparser. Crops failed, famine spread. Mankind began to undergo great sufferings; daughters hid food from their mothers, mothers ate their young. At the urging of Enlil, the gods refrained from helping Mankind: Let them starve, let them be decimated, Enlil decreed.

In the "Great Below"—in Antarctica—the Ice Age was also causing changes. From year to year, the ice cap covering the continent at the South Pole grew thicker and thicker. Under the increasing pressure of its weight, friction and heat increased at its bottom. Soon the immense ice cap was floating on a slippery slush of mud. From the orbiting shuttlecraft, an alarm was sounded: the ice cap was becoming unstable; if it should happen to slip off the continent into the ocean—an immense tidal wave would engulf all of Earth!

It was not an idle danger. In the heavens, the Twelfth Planet was orbiting back to the Place of Crossing between Jupiter and Mars. As on previous occasions when it neared Earth, its gravitational pull caused earthquakes and other disturbances upon Earth and in its celestial motions. Now, it was calculated, this gravitational pull could well trigger the slippage of the ice cap, and inundate Earth with a global deluge. From this catastrophe, the astronauts themselves could not be immune.

As preparations were made to assemble all the Anunnaki near the Spaceport, and ready the craft to take them aloft before the tidal wave struck, ruses were employed to keep the approaching catastrophe a secret from Mankind. Fearing that the Spaceport would be mobbed, all the gods were made to swear to secrecy. And as to Mankind, Enlil said—Let them perish; let the seed of Earthling be wiped off the face of Earth.

In Shuruppak, the city under the lordship of Ninhursag, relations between Man and gods had gone the farthest. There, for the first time ever, a man was elevated to the status of king. As the sufferings of Mankind increased, ZI.U.SUD.RA (as the Sumerians called him) pleaded for the help of Ea. From time to time, Ea and his seafarers clandestinely brought Ziusudra and his people a load of fish. But now the question involved the very destiny of Mankind. Shall all the handiwork of Ea and Ninhursag perish "and turn to clay" as Enlil wished—or should the Seed of Mankind be preserved?

Acting on his own, but mindful of his oath, Ea saw in Ziusudra the chance to save Mankind. The next time Ziusudra came to pray and plead in the temple, Ea began to whisper from behind a screen. Pretending to talk to himself, Ea gave Ziusudra urgent instructions:

> Tear down the house, build a ship!
> Give up possessions, seek thou life!
> Foreswear belongings, keep soul alive!
> Aboard ship take thou the seed of all living things.
> That ship thou shalt build;
> Her dimensions shall be to measure.

The ship was to be a submersible vessel, a "submarine" that could withstand the avalanche of water. The Sumerian texts contain the dimensions and other structural instructions for the various decks and compartments in such detail that it is possible to draw the ship, as was done by Paul Haupt (Fig. 56). Ea also provided Ziusudra with a navigator, instructing him to direct the vessel toward the "Mount of Salvation," Mount Ararat; as the highest range in the Near East, its peaks would be the first to emerge from under the waters.

The Deluge came as expected. "Gathering speed as it

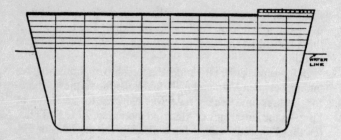

Fig. 56

blew" from the south, "submerging the mountains, over-taking the people like a battle." Viewing the catastrophe from above, as they orbited Earth in their craft, the Anunnaki and their leaders realized how much they had fallen in love with Earth and with Mankind. "Ninhursag wept . . . the gods wept with her for the land. . . . The Anunnaki, all humbled, sit and weep" as they huddled, cold and hungry, in their Shuttlecraft.

When the waters subsided and the Anunnaki began to land on Ararat, they were elated to discover that the Seed of Mankind was saved. But as Enlil too arrived, he was furious to see that "a living soul had escaped." It took the pleadings of the Anunnaki and the persuasion of Ea, to bring him around to their point of view—that if Earth was to be resettled, the services of Man were indispensable.

And so it was, that the sons of Ziusudra and their families were sent out to settle the mountain ranges flanking the plain of the two rivers, to await the time when the plain was dry enough to inhabit. As to Ziusudra, the Anunnaki

> Life like that of a god they gave him;
> Breath eternal, like a god, they granted him.

This they achieved by exchanging his "Breath of Earth" with the "Breath of Heaven." Then they took Ziusudra, "the preserver of the seed of Mankind," and his wife, to "reside in the faraway place"—

> In the Land of the Crossing,
> The Land *Tilmun,*
> The place where Utu rises,
> They caused him to dwell.

It is evident by now that the Sumerian tales of the Gods of Heaven and Earth, of the Creation of Man and of the Deluge, were the fountainhead from which the other nations of the ancient Near East drew their knowledge, beliefs and "myths." We have seen how the Egyptian beliefs matched the Sumerian ones, how their first sacred city was named after An, how the *Ben-Ben* resembled the Sumerian GIR, and so on.

It is also generally accepted by now, that the biblical tales of the Creation and of the events leading to the Deluge are condensed Hebrew versions of the Sumerian traditions. The biblical hero of the Deluge, Noah, was the equivalent of the Sumerian Ziusudra (who was called *Utnapishtim* in the Akkadian versions). But while the Sumerians asserted that the hero of the Deluge was made immortal, no such claim is made in the Bible for Noah. The immortalization of Enoch is also given short shrift, quite unlike the detailed Sumerian tale of Adapa or other texts dealing with other Ascents. But this abrupt biblical attitude could not prevent the spread, over the millenia, of legends dealing with the biblical heroes and their sojourn in, or return to, Paradise.

According to very ancient legends, which survived in a number of versions stemming from a composition almost 2,000 years old called *The Book of Adam and Eve,* Adam fell sick after he was 930 years old. Seeing his father "sick and in pain," his son Seth volunteered to go to "the nearest gate of Paradise . . . and lament and make entreaty to God; perchance He will hearken to me and send His angel to bring me the fruit, for which thou hast longed"—the fruit of the Tree of Life.

But Adam, accepting his mortal's fate, only wished for the excruciating pains to be relieved. So he asked Eve his wife to take Seth, and together go "to the neighborhood of Paradise"; there to ask not for the Fruit of Life, but only for one drop of the "oil of life" which floweth from the Tree, "to anoint me with it, that I may have rest from these pains."

Having done as Adam asked, Eve and Seth reached the gates of Paradise, and entreated with the Lord. Finally, the angel Michael appeared unto them—only to announce that their request would not be granted. "The time of Adam's life is fulfilled," the angel said; his death was not to be averted or postponed. Six days later, Adam died.

Even the historians of Alexander created a direct link between his miraculous adventures and Adam, the very first man who had dwelt in Paradise, and was proof of its existence and lifegiving powers. The connecting link in the case of Alexander was the unique stone which emitted light: it was said to have been brought out of the Garden of Eden by Adam, then handed down from generation to generation, until it reached the hands of an immortal Pharaoh, who in turn gave it to Alexander.

The plot-of-parallels indeed thickens, as one realizes that there exists an old Jewish legend, whereby the staff with which Moses performed many miracles, including the parting of the waters of the Lake of Reeds, was brought out of the Garden of Eden by Adam. Adam gave it to Enoch; Enoch gave it to his great-grandson Noah, the hero of the Deluge. Then it was handed down through the line of Shem, son of Noah, from generation to generation, until it reached Abraham (the first Hebrew patriarch). Abraham's great-grandson Joseph brought it with him to Egypt, where he rose to highest rank in the Pharaoh's court. There the staff remained among the treasures of the Egyptian kings; and thus it reached the hands of Moses, who was raised as an Egyptian prince before he escaped into the Sinai peninsula. In one version, the staff was carved out of a single stone; in another, it was made of a branch of the Tree of Life, which grew in the Garden of Eden.

In these interwoven relationships, harkening back to the earliest times, there were also tales linking Moses with Enoch. A Jewish legend, called "The Ascent of Moses," relates that when the Lord summoned Moses at Mount Sinai and charged him to lead the Israelites out of Egypt, Moses resisted the mission for various reasons, including his slow and non-eloquent speech. Determined to remove his meekness, the Lord decided to show him His throne and "the angels of the Heavens" and the mysteries thereof. So "God commanded Metatron, the Angel of the Countenance, to conduct Moses to the celestial regions." Terrified, Moses asked Metatron: "Who art thou?" And the angel (literally: "emissary") of the Lord replied: "I am Enoch, son of Jared, thy ancestor." (Accompanied by the angelic Enoch, Moses soared through the Seven Heavens, and saw Hell and Paradise; then he was returned to Mount Sinai, and accepted his mission.)

Further light on the occurrences concerning Enoch, and his preoccupation with the impending Deluge and its hero, his great-grandson Noah, is shed by yet another ancient book, the *Book of Jubilees*. It was also known in early times as the *Apocalypse of Moses*, for it allegedly was written down by Moses at Mount Sinai as an angel dictated to him the histories of days past. (Scholars, though, believe that the work was composed in the second century B.C.)

It follows closely the biblical narratives of the Book of Genesis; yet it provides more detail, such as the names of wives and daughters of the pre-Diluvial patriarchs. It also enlarges upon the events experienced by Mankind in those prehistoric days. The Bible informs us that the father of Enoch was *Jared* ("Descent"), but not why he was so named. The *Book of Jubilees* provides the missing information. It says that the parents of Jared so named him,

For in his days the angels of the Lord descended upon Earth—those who are named *The Watchers*—that they

should instruct the children of men, that they should
do judgment and uprightness upon Earth.

Dividing the eras into "jubilees," the *Book of Jubilees*
further narrates that "in the eleventh jubilee Jared took to
himself a wife; her name was *Baraka* ("Lightning Bright"),
the daughter of Rasujal, a daughter of his father's brother . . .
and she bare him a son and he called his name Enoch. And
he (Enoch) was the first among men that are born on Earth
who learnt writing and knowledge and wisdom, and who
wrote down the signs of heaven according to the order of
their months in a book, that men might know the seasons of
the year according to the order of their separate mouths."

In the twelfth jubilee, Enoch took as wife *Edni* ("My
Eden"), the daughter of Dan-el. She bare him a son whose
name was Methuselah. After that, Enoch "was with the angels of God for six jubilees of years, and they showed him
everything which is on Earth and in the Heavens . . . and he
wrote down everything."

But by then, trouble was brewing. The Book of Genesis
reports that it was before the Deluge, "That the sons of the
gods saw the daughters of Man, that they were good, and
they took them for wives of all which they chose . . . and it
repented the Lord that He had made Man on Earth . . . and
the Lord said: I will destroy the Man whom I had created
from the face of the Earth."

According to the *Book of Jubilees,* Enoch played some
role in this changed attitude by the Lord, for "he testified
about the Watchers who had sinned with the daughters of
men; he testified against them all." And it was to protect him
from the revenge of the sinning Angels of the Lord, that "he
was taken from amongst the children of men, and was conducted into the Garden of Eden." Specifically named as one
of the four places of God on Earth, it was in the Garden of
Eden that Enoch was hidden, and where he wrote down his
Testament.

It was after that that Noah, the righteous man singled out

to survive the Deluge, was born. His birth, occurring at the troubled times when the "sons of the gods" were indulging in sex with mortal females, caused a marital crisis in the patriarchal family. As the Book of Enoch tells it, Methuselah "took a wife for his son Lamech, and she became pregnant by him and bore a son." But when the baby—Noah—was born, things were not as usual:

> His body was white as snow and red as the blooming of a rose, and the hair of his head and his long locks were white as wool, and his eyes were beautiful.
>
> And when he opened his eyes, he lighted up the whole house like the sun, and the whole house was very bright.
>
> And thereupon he arose in the hands of the midwife, opened his mouth, and conversed with the Lord of Righteousness.

Shocked, Lamech ran to his father Methuselah, and said:

> I have begotten a strange son, diverse from and unlike Man, and resembling the sons of the God of Heaven; and his nature is different, and he is not like us . . .
>
> And it seems to me that he is not sprung from me but from the angels.

Suspecting, in other words, that his wife's pregnancy was induced not by him, but that she was impregnated by one of the angels, Lamech had an idea: Since his grandfather Enoch was staying amongst the Sons of the Gods, why not ask him to get to the bottom of this? "And now, my father," he said to Methuselah, "I petition thee and implore thee that thou mayest go to Enoch thy father, and learn from him the truth, for his dwelling place is amongst the angels."

Methuselah went as Lamech had asked, and reaching the Divine Abode summoned Enoch, and reported the unusual baby boy. Making some inquiries, Enoch assured Methuselah that Noah was indeed a true son of Lamech; and that his unusual countenance was a sign of things to come: "There shall be a Deluge and great destruction for one year," and only this son, who is to be named *Noah* ("Respite") and his family shall be saved. These future things, Enoch told his son, "I have read in the heavenly tablets."

The term employed in these ancient, even if ex-biblical scriptures, to denote the "sons of the gods" involved in the pre-Diluvial shenanigans, is "*Watchers.*" It is the very term *Neter* ("Watchers") by which the Egyptians called the gods, and the exact meaning of the name *Shumer,* their landing place on Earth.

The various ancient books which throw this extra light on the dramatic events in the days before the Deluge, have been preserved in several versions that are all only translations (direct and indirect) of lost Hebrew originals. Yet their authenticity was confirmed with the renowned discoveries in recent decades of the Dead Sea Scrolls, for among the finds were fragments of scrolls which were undoubtedly parts of the Hebrew originals of such "memoirs of the Patriarchs."

Of particular interest to us is a scroll fragment which deals with the unusual birth of Noah, and from which we can learn the original Hebrew term for what has been translated as "Watchers" or "Giants," not only in the ancient versions, but even by modern scholars (as T. H. Gaster, *The Dead Sea Scriptures* and H. Dupont-Sommer, *The Essene Writings from Qumran*). According to these scholars, column II of the scroll fragment begins thus:

> Behold, I thought in my heart that the conception was from one of the *Watchers,* one of the Holy Ones, and (that the child really belonged) to the *Giants.*
>
> And my heart was changed within me because of the child.
>
> Then I, Lamech, hastened and went to Bath-Enosh (my) wife, and I said to Her:

[I want you to take an oath] by the Most High, by
the Lord Supreme, the King of all the worlds,
    the ruler of the Sons of Heaven, that you will tell me
in truth whether . . .

But as we examine the Hebrew original (Fig. 57) we find
that it does not say "Watchers"; it says *Nefilim*—the very
term used in Genesis 6.

Thus do all the ancient texts and all the ancient tales confirm each other:

The days before the Deluge were the days when "The
Nefilim were upon the Earth—the Mighty Ones, the People
of the Rocketships."

Column II

1. הא באדן חשבת בלבי די מן עירין הריאנתא ומן קדישין הויא ‎ולנפילין

2. ולבי עלי משתני על עולימא דנא
3. באדן אנה למך אתבהלת ועלת על בתאנוש אנותתי ואמרת

4.] ‏אנא ועד בעליא במרה רבותא במלך כול עולמים

Fig. 57

In the words of the Sumerian King Lists, "the Deluge has
swept over" 120 *shars*—120 orbits of 3,600 years each—
after the first landing on Earth. This places the Deluge at
about 13,000 years ago. It is exactly the time when the last
ice age ended abruptly, when agriculture began. It was followed 3,600 years later by the New Stone Age (as scholars
call it), the age of pottery. Then, 3,600 years later, Civilization all at once blossomed out—in the "plain between the
rivers," in Shumer.

"And the whole Earth was of one language and of one
kind of things," the Book of Genesis says; but soon after the
people had established themselves in the Land of Shin'ar

(Sumer), and built dwellings of fired clay bricks, they conspired to "build a city, and a Tower the top of which can reach unto Heaven."

The Sumerian texts from which this biblical tale was extracted have not yet been found; but we do come across allusions to the event in various Sumerian tales. What emerges is an apparent effort on the part of Ea to enlist Mankind in gaining control over the space facilities of the Nefilim—one more incident in the continuing feud between Ea and Enlil, which by then had spilled over to their offspring. As a result of the incident, the Bible tells us, the Lord and his unnamed colleagues decided to disperse Mankind and "confuse" its languages—give it diverse and separate civilizations.

The deliberations of the gods in the era following the Deluge are mentioned in various Sumerian texts. The one called the Epic of Etana states:

The Great Anunnaki who decree the fate sat exchanging their counsels regarding the Earth. They who created the four regions, who set up the settlements, who oversaw the land, were too lofty for Mankind.

The decision to establish on Earth four Regions was thus coupled with a decision to install intermediaries (priest-kings) between the gods and Mankind; so "kingship was again lowered to Earth from Heaven."

In an effort—which proved futile—to end or abate the feud between the Enlil and Ea families, lots were drawn between the gods to determine who would have dominion over which of the Regions. As a result, Asia and Europe were assigned to Enlil and his offspring; to Ea, Africa was given.

The First Region of civilization was Mesopotamia and the lands bordering upon it. The mountain-lands where agriculture and settled life began, the lands that came to be known as Elam, Persia, Assyria—were given to Enlil's son NIN.UR.TA, his rightful heir and "Foremost Warrior." Some Sumerian texts have been found dealing with Ninurta's heroic efforts to dam the mountain passes and assure the survival of his human subjects in the harsh times that had followed the Deluge.

When the layers of mud that had covered the Plain between the Two Rivers dried up sufficiently to permit re-settlement, Shumer and the lands that stretched therefrom westward to the Mediterranean were put under the charge of Enlil's son NAN.NAR (*Sin* in Akkadian). A benevolent god, he supervised the reconstruction of Sumer, rebuilding pre-Diluvial cities at their original sites and establishing new cities. Among the latter was his favorite capital *Ur,* the birthplace of Abraham. His depictions included the crescent symbol of the Moon, which was his celestial "counterpart" (Fig. 58). To Enlil's youngest son, ISH.KUR (whom the Akkadians called *Adad*), were given the northwestern lands, Asia Minor and the Mediterranean islands from where civilization—"Kingship"—eventually spread to Greece. Like Zeus in later Greece, Adad was depicted riding a bull and holding a forked lightning.

Ea too divided the Second Region, Africa, among his sons. It is known that a son named NER.GAL lorded over the southernmost parts of Africa. A son named GI.BIL learned from his father the arts of mining and metallurgy, and took over control of the African gold mines. A third son—Ea's favorite—was named by him after the home planet *MARDUK,* and was taught by Ea all knowledge of sciences and astronomy. (Circa 2000 B.C., Marduk usurped the Lordship of Earth and was declared Supreme God of Babylon and of "the Four Quarters of the Earth.") And, as we have seen, a son whose Egyptian name was *Ra* presided over the core civilization of this Region, the civilization of the Nile Valley.

The Third Region, as was discovered only some fifty years ago, was in the subcontinent of India. There too, a great civilization arose in antiquity, some 1,000 years after the Sumerian one. It is called the Indus Valley Civilization, and its center was a royal city unearthed at a site called *Harappa.* Its people paid homage not to a god but to a goddess, depict-

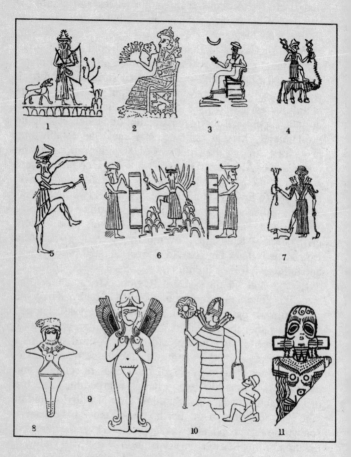

THE GODS OF THE EARTH

1. ENLIL 2. NINURTA 3. NANNAR/Sin 4. ISHKUR/Adad 5. NERGAL 6. GIBIL
7. MARDUK. IRNINI/Ishtar as Great lady (8), Enchantress (9), Warrior (10),
Pilot (11)

Fig. 58

ing her in clay figurines as an enticing female, adorned with necklaces, her breasts enhanced by straps which crossed her body.

Because the script of the Indus Civilization is still undeciphered, no one knows what the Harappans called their goddess, or who exactly she was. It is our conclusion, however, that she was the daughter of Sin, whom the Sumerians called IR.NI.NI ("The Strong, Sweetsmelling Lady") and the Akkadians called *Ishtar.* Sumerian texts tell of her dominion in a far land named *Aratta*—a land of grain crops and granaries as Harappa was—whereto she made flying trips, attired as a pilot.

It was in need of a Spaceport that the Fourth Region was set aside by the Great Anunnaki—a Region not for Mankind, but for their own exclusive use. All their space facilities from the time they had landed on Earth—the Spaceport at Sippar, the Mission Control Center at Nippur—were wiped out by the Deluge. The low-lying Mesopotamian plain was still too muddy for millenia to enable the rebuilding there of these vital installations. Another place—more elevated yet suitable, secluded but accessible—had to be found for the Spaceport and its auxiliary installations. It was to be a "sacred zone"—a restricted area, accessible only by permission. It was called in Sumerian TIL.MUN—literally, "Land of the Missiles."

In charge of this post-Diluvial Spaceport they put the son of Sin (and thus a grandson of Enlil), a twin brother of Irnini/Ishtar. His name was UTU ("The Bright One")—*Shamash* in Akkadian. It was he who ably carried out Operation Deluge—the evacuation out of Sippar. He was the chief of the Spacemen based on Earth, the "Eagles"; and he proudly wore his Eagle-uniform on formal occasions (Fig. 59).

In the days before the Deluge, traditions held, a few chosen mortals had been taken aloft from the Spaceport: *Adapa,* who missed his chance; and *Enmeduranki,* whom the gods Shamash and Adad transported to the Celestial Abode, to be initiated in priestly secrets (and then returned to Earth). Then there was *Ziusudra* ("His Life-Days Prolonged"), hero of the Deluge, who was taken with his wife to live in Tilmun.

In post-Diluvial times, Sumerian records stated, *Etana*—an early ruler of Kish—was taken aloft in a *Shem* to the Abode of the Gods, there to be granted the Plant of Rejuvenation and Birthgiving (but he was too frightened to complete the journey). And the Pharaoh Thothmes III claimed in his inscriptions that the god Ra had taken him aloft, given him a tour of the heavens, and returned him to Earth:

> He opened for me the doors of Heaven,
> He spread open for me the portals of its horizon.
> I flew up to the sky as a divine Falcon . . .
> That I might see his mysterious ways in heaven . . .
> I was made full with the understanding of the gods.

In the later memories of Mankind, the *Shem* was cherished as an obelisk, and the rocketship saluted by "Eagles" gave way to a sacred Tree of Life (Fig. 60). But in Sumer, where the gods were a present reality—as in Egypt when the first Pharaohs had reigned—Tilmun, the "Land of the Missiles," was a real place: a place where Man could find Immortality.

And there, in Sumer, they recorded the tale of a man who—uninvited by the gods—set out to reverse his fate nevertheless.

Fig. 59

Fig. 60

VII

Gilgamesh: The King Who Refused to Die

The Sumerian tale of the first known Search for Immortality concerns a ruler of long long ago, who asked his divine godfather to let him enter the "Land of the Living." Of this unusual ruler, ancient scribes wrote down epic tales. They said of him that

> Secret things he has seen;
> What is hidden from Man, he found out.
> He even brought tidings
>   of the time before the Deluge;
> He also took the distant journey,
>   wearisome and under difficulties.
> He returned, and upon a stone column
>   all his toil he engraved.

Of that olden Sumerian tale, less than two hundred lines have remained. Yet we know it from its translations into the languages of the peoples who followed the Sumerians in the Near East: Assyrians, Babylonians, Hittites, Hurrians. They all told and retold the tale; and the clay tablets on which these later versions were written down—some intact, some damaged, many fragmented beyond legibility—have enabled many scholars over the better part of a century to piece the tale together.

At the core of our knowledge are twelve tablets in the Akkadian language; they were part of the library of Ashurbani-pal in Nineveh. They were first reported by George Smith, whose job at the British Museum in London was to sort out, match and categorize the tens of thousands of tablets and tablet fragments that arrived at the Museum from Mesopotamia. One day, his eye caught a fragmented text which appeared to relate the story of the Deluge. There was no mistaking: the cuneiform texts, from Assyria, were telling of a king who sought out the hero of the Deluge, and heard from him a first-person account of the event!

With understandable excitement, the Museum directors sent George Smith to the archaeological site to search for missing fragments. With luck, he found enough of them to be able to reconstruct the text and guess the sequence of the tablets. In 1876, he conclusively showed that this was, as his work was titled, *The Chaldean Account of the Flood.* From the language and style he concluded that it "was composed in Babylon circa 2000 B.C."

George Smith at first read the name of the king who searched for Noah *Izdubur,* and suggested that he was none other than the biblical hero-king Nimrod. For a time scholars believed that the tale indeed concerned the very first mighty king, and referred to the twelve-tablet text as the "Nimrod Epos." More finds and much further research established the Sumerian origin of the tale, and the true reading of the hero's name: GIL.GA.MESH. It has been confirmed from other historical texts—including the Sumerian King Lists—that he was a ruler of Uruk, the biblical Erech, circa 2900 B.C. *The Epic of Gilgamesh,* as this ancient literary work is now called, thus takes us back nearly 5,000 years.

One must understand the history of Uruk to grasp the Epic's dramatic scope. Affirming the biblical statements, the Sumerian historical records also reported that in the aftermath of the Deluge, kingship—royal dynasties—indeed began at Kish; it then was transferred to Uruk as a result of the ambitions of Irnini/Ishtar, who cherished not at all her domain far away from Sumer.

Uruk, initially, was only the location of a sacred precinct, where an Abode (temple) for An, the "Lord of Heaven," was perched atop a vast ziggurat named E.AN.NA ("House of An"). On the rare occasion of An's visits to Earth, he took a liking to Irnini. He bestowed on her the title IN.AN.NA— "Beloved of An" (the ancient gossip suggested that she was beloved in more than platonic ways), and installed her in the Eanna, which otherwise stood unoccupied.

But what good was a city without people, a lordship with no one to rule over? Not too far away to the south, on the shores of the Persian Gulf, Ea lived in Eridu in semi-isolation. There he kept track of human affairs, dispensing knowledge and civilization to mankind. Enchanting and perfumed, Inanna paid Ea (a great-uncle of hers) a visit. Enamored and drunk, Ea granted her wish: to make Uruk the new center of Sumerian civilization, the seat of kingship in lieu of Kish.

To carry out her grandiose plans, whose ultimate goal was to break into the Inner Circle of the Twelve Great Gods, Inanna-Ishtar enlisted the support of her brother Utu/ Shamash. Whereas in the days before the Deluge the intermarriage between the Nefilim and the daughters of Man brought about the wrath of the gods, the practice was no longer frowned upon in the aftermath of the Deluge. And so it was, that the high priest at the temple of An was at the time a son of Shamash by a human female. Ishtar and Shamash anointed him as king of Uruk, starting the world's first dynasty of priestly kings. According to the Sumerian King Lists, he ruled for 324 years. His son, "who built Uruk," ruled for 420 years. When Gilgamesh, the fifth ruler of this dynasty, ascended the throne, Uruk was already a thriving Sumerian center, lording over its neighbors and trading with far lands. (Fig. 61).

An offspring of the great god Shamash on his father's side, Gilgamesh was considered to be "two-thirds god, one-third human" by the further fact that his mother was the goddess NIN.SUN (Fig. 62). He was thus accorded the privilege of having his name written with the prefix "divine."

Fig. 61

Fig. 62

Proud and self-assured, Gilgamesh began as a benevolent and conscientious king, engaged in the customary tasks of raising the city's ramparts or embellishing the temple precinct. But the more knowledge he acquired of the histories of gods and men, the more he became philosophical and restless. In the midst of merriment, his thoughts turned to death. Would he, by virtue of his divine two-thirds, live as long as his demi-god forefathers—or would his one-third prevail, and determine for him the life span of a mortal human? Before long, he confessed his anxiety to Shamash:

> In my city man dies; oppressed is my heart.
> Man perishes; heavy is my heart . . .
> Man, the tallest, cannot stretch to heaven;
> Man, the widest, cannot cover the earth.

"Will I too 'peer over the wall'?" he asked Shamash; "will I too be fated thus?"

Evading a direct answer—perhaps not knowing it himself—Shamash attempted to have Gilgamesh accept his fate, whatever it might be, and to enjoy life while he could:

> When the gods created Mankind,
> Death for Mankind they allotted;
> Life they retained in their own keeping.

Therefore, said Shamash,

> Let full be thy belly, Gilgamesh;
> Make thou merry by day and night!
> Of each day, make thou a feast of rejoicing;
> Day and night, dance thou and play!
> Let thy garments be sparkling fresh,
>   thy head washed; bathe thou in water.
> Pay heed to the little one that holds thy hand,
>   let thy spouse delight in thy bosom;
>   for this is the fate of Mankind.

But Gilgamesh refused to accept this fate. Was he not two-thirds divine, and only one-third human? Why then should the lesser mortal part, rather than his greater godly element, determine his fate? Roving by daytime, restless at night, Gilgamesh sought to stay young by intruding on newlywed couples and insisting on having intercourse with the bride ahead of the bridegroom. Then, one night, he saw a vision which he felt was an omen. He rushed to his mother to tell her what he saw, so that she might interpret the omen for him:

> My mother.
> During the night, having become lusty,
> I wandered about.
> In the midst (of night) omens appeared.
> A star grew larger and larger in the sky.
> The handiwork of Anu descended towards me!

"The handiwork of Anu" that descended from the skies
fell to Earth near him, Gilgamesh continued to relate:

> I sought to lift it;
>   it was too heavy for me.
> I sought to shake it;
> I could neither move nor raise it.

While he was attempting to shake loose the object,
which must have embedded itself deep into the ground,
"the populace jostled toward it, the nobles thronged about
it." The object's fall to Earth was apparently seen by many,
for "the whole of Uruk land was gathered around it." The
"heroes"—the strongmen—then lent Gilgamesh a hand in
his efforts to dislodge the object that fell from the skies:
"The heroes grabbed its lower part, I pulled it up by its
forepart."

While the object is not fully described in the texts, it was
certainly not a shapeless meteor, but a crafted object worthy
of being called the *handiwork* of the great Anu himself. The
ancient reader, apparently, required no elaboration, having
been familiar with the term ("Handiwork of Anu") or with its
depiction, as possibly the one shown on an ancient cylinder
seal (Fig. 63).

The Gilgamesh text describes the lower part, which was
grabbed by the heroes, by a term that may be translated
"legs." It had, however, other pronounced parts and could
even be entered, as becomes clear from the further description by Gilgamesh of the night's events:

> I pressed strongly its upper part;
> I could neither remove its covering,
>   nor raise its Ascender . . .
> With a destroying fire its top I (then) broke off,
>   and moved into its depths.
> Its movable That Which Pulls Forward
> I lifted, and brought it to thee.

Gilgamesh was certain that the appearance of the object was an omen from the gods concerning his fate. But his mother, the goddess Ninsun, had to disappoint him. That which descended like a star from Heaven, she said, foretells the arrival of "a stout comrade who rescues; a friend is come to thee . . . he is the mightiest in the land . . . he will never forsake thee. This is the meaning of thy vision."

She knew what she was talking about; for unbeknown to Gilgamesh, in response to pleas from the people of Uruk that something be done to divert the restless Gilgamesh, the gods arranged for a wild man to come to Uruk and engage Gilgamesh in wrestling matches. He was called ENKI.DU—"A Creature of Enki"—a kind of Stone Age Man who had been living in the wilderness among the animals and as one of them: "The milk of wild creatures he was wont to suck." He was depicted naked, bearded, with shaggy hair—often shown in the company of his animal friends (Fig. 64).

To tame him, the nobles of Uruk assigned a harlot. Enkidu, until then knowing only the company of animals, regained his human element as he made love to the woman, over and over again. Then she brought Enkidu to a camp outside town, where he was coached in the speech and manners of Uruk and in the habits of Gilgamesh. "Restrain Gilgamesh, be a match for him!" the nobles told Enkidu.

The first encounter took place at night, as Gilgamesh left his palace and started to roam the streets, looking for sexual adventures. Enkidu met him in the street, barring his way. "They grappled each other, holding fast like bulls." Walls shook, doorposts were shattered as the two wrestled. At last, "Gilgamesh bent the knee"; the match was over: He lost to the stranger. "His fury abated, Gilgamesh turned away." Just then, Enkidu addressed him, and Gilgamesh recalled his mother's words. Here then was his new "stout friend." "They kissed each other, and formed a friendship."

As the two became inseparable friends, Gilgamesh began

Fig. 63

Fig. 64

to reveal to Enkidu his fear of a mortal's fate. On hearing this, "the eyes of Enkidu filled with tears, ill was his heart, bitterly he sighed." Then he told Gilgamesh, that there is a way to outsmart his fate: to force his way into the secret Abode of the Gods. There, if Shamash and Adad would stand by him, the gods could accord him the divine status to which he was entitled.

The "Abode of the Gods," Enkidu related, was in "the cedar mountain." He happened to discover it, he said, as he was roaming the lands with the wild beasts; but it was guarded by a fearsome monster named *Huwawa:*

> I found it, my friend, in the mountains
>   as I was roaming with the wild beasts.
> For many leagues extends the forest:
> I went down into its midst.
> Huwawa (is there); his roar is like a flood,
>   his mouth is fire,
>   his breath is death . . .

The Cedar Forest's watcher, the Fiery Warrior,
   is mighty, never resting . . .
To safeguard the Cedar Forest,
   as a terror to mortals the god Enlil appointed him.

The very fact that Huwawa's main duty was to prevent
mortals from entering the Cedar Forest only whetted the determination of Gilgamesh to reach the place; for surely, it
was there that he could join the gods and escape his mortal's
fate:

Who, my friend, can scale heaven?
Only the gods,
   by going to the underground place of Shamash.
Mankind's days are numbered;
   whatever they achieve is but the wind.
Even thou art afraid of death,
   in spite of your heroic might.
Therefore,
Let me go ahead of thee,
   let thy mouth call to me:
"Advance, fear not!"

This, then, was the plan: by going to "the underground
place of Shamash" in the Cedar Mountain, to be enabled to
"scale heaven" as the gods do. Even the tallest man, Gilgamesh earlier pointed out, "cannot stretch to heaven." Now
he knew where the place was, from which Heaven could be
scaled. He fell to his knees and prayed to Shamash: "Let
me go, O Shamash! My hands are raised in prayer . . . to
the Landing Place, give command . . . Establish over me thy
protection!"

The text's lines containing the answer of Shamash are,
unfortunately, broken off the tablet. We do learn that "when
Gilgamesh inspected his omen . . . tears ran down his face."
Apparently he was permitted to go ahead—but at his own
risk. Nevertheless, Gilgamesh decided to proceed, and

fight Huwawa without the god's aid. "Should I fail," he said, people will remember me: "Gilgamesh, they will say, against fierce Huwawa has fallen." But should I succeed, he continued—I will obtain a *Shem*—the vehicle "by which one attains eternity."

As Gilgamesh ordered special weapons with which to fight Huwawa, the elders of Uruk tried to dissuade him. "Thou are yet young, Gilgamesh," they pointed out—and why risk death with so many sure years to live anyway, against unknown odds of success: "That which thou wouldst achieve, thou knowest not." Gathering all available information about the Cedar Forest and its guardian, they cautioned Gilgamesh:

> We hear that Huwawa is wondrously built;
> Who is there to face his weapons?
> Unequal struggle it is
>   with the siege-engine Huwawa.

But Gilgamesh only "looked around, smiling at his friend." The talk of Huwawa as a mechanical monster, a "siege engine" that is "wondrously built," only encouraged him to believe that it was indeed controllable by commands from the gods Shamash and Adad. Since he himself did not succeed in obtaining a clear-cut promise of support from Shamash, Gilgamesh decided to enlist his mother in the effort. "Grasping each other, hand in hand, Gilgamesh and Enkidu to the Great Palace go, to the presence of Ninsun, the Great Queen. Gilgamesh came forward as he entered the palace: 'O Ninsun (he said) . . . a far journey I have boldly undertaken, to the place of Huwawa; an uncertain battle I am about to face; unknown pathways I am about to ride. Oh my mother, pray thou to Shamash on my behalf!'"

Obliging, "Ninsun entered her chamber, put on a garment as beseems her body, put on an ornament as beseems her breast . . . donned her tiara." Then she raised her hands in prayer to Shamash—putting the onus of the voyage on him:

"Why," she asked rhetorically, "having given me Gilgamesh for a son, with a restless heart didst thou endow him? And now, thou didst affect him to go on a far journey, to the place of Huwawa!" She called upon Shamash to protect Gilgamesh:

> Until he reaches the Cedar Forest,
> Until he has slain the fierce Huwawa,
> Until the day that he goes and returns.

As the populace heard that Gilgamesh was going to "the Landing Place" after all, "they pressed closer to him" and wished him success. The city elders offered more practical advice: "Let Enkidu go before thee: he knows the way . . . in the forest, the passes of Huwawa let him penetrate . . . he who goes in front saves the companion!" They too invoked the blessings of Shamash: "Let Shamash grant thee thy desire; what thy mouth hath spoken, let him show thine eyes; may he open for thee the barred path, the road unclose for thy treading, the mountain unclose for thy foot!"

Ninsun had a few parting words. Turning to Enkidu, she asked him to protect Gilgamesh; "although not of my womb's issue art thou, I herewith adopt thee (as a son)," she told him; guard the king as thy brother! Then she placed her emblem around the neck of Enkidu.

And the two were off on their dangerous quest.

The fourth tablet of the Epic of Gilgamesh is devoted to the comrades' journey to the Cedar Forest; unfortunately, the tablet is so fragmented that, in spite of the discovery of parallel fragments in the Hittite language, no cohesive text could be put together.

It is evident, however, that they traveled a great distance, toward a western destination. On and off, Enkidu tried to persuade Gilgamesh to call off the quest. Huwawa, he said, can hear a cow moving sixty leagues away. His "net" can grasp from great distances; his call reverberates from the "Place Where the Rising Is Made" as far back as to Nippur; "weakness lays hold on him" who approaches the forest's gates. Let us turn back, he pleaded. But proceed they did:

> At the green mountain the two arrived.
> Their words were silenced;
> They themselves stood still.
> They stood still and gazed at the forest;
> They looked at the height of the cedars;
> They looked at the entrance to the forest.
> Where Huwawa wont to move was a path:
> > straight were the tracks, a fiery channel.
> They beheld the Cedar Mountain,
> Abode of the Gods,
> > the Crossroads of Ishtar.

Awestruck and tired, the two lay down to sleep. In the middle of the night they were awakened. "Didst thou arouse me?" Gilgamesh asked Enkidu. No, said Enkidu. No sooner had they dozed off than Gilgamesh again awakened Enkidu. He had witnessed an awesome sight, he said—unsure whether he was awake or dreaming:

> In my vision, my friend,
> > the high ground toppled.
> It laid me low, trapped my feet . . .
> The glare was overpowering!
> A man appeared;
> > the fairest in the land was he . . .
> From under the toppled ground he pulled me out.
> He gave me water to drink; my heart quieted.
> On the ground he set my feet.

Who was this "man"—"the fairest in the land"—who pulled Gilgamesh from under the toppled ground? What was the "overpowering glare" that accompanied the landslide?

Enkidu had no answers; tired, he went back to sleep. But the night's tranquility was shattered once again:

> In the middle of the watch,
>> the sleep of Gilgamesh was ended.
> He started up, saying to his friend:
> "My friend, didst thou call me?
> Why am I awake?
> Didst thou not touch me?
> Why am I startled?
> Did not some god go by?
> Why is my flesh numb?"

Denying that he had awakened Gilgamesh, Enkidu left his comrade wondering whether it was "some god who went by." Bewildered, the two fell asleep again, only to be awakened once more. This is how Gilgamesh described what he saw:

> The vision that I saw was wholly awesome!
> The heavens shrieked, the earth boomed.
> Though daylight was dawning, darkness came.
> Lightning flashed, a flame shot up.
> The clouds swelled; it rained death!
> Then the glow vanished; the fire went out.
> And all that had fallen was turned to ashes.

Gilgamesh must have realized that he had witnessed the ascent of a "Sky Chamber": the shaking ground as the engines ignited and roared; the clouds of smoke and dust that enveloped the site, darkening the dawn sky; the brilliance of the engines' fire, seen through the thick clouds; and—as the jetcraft was aloft—its vanishing glow. A "wholly awesome" sight indeed! But one which only encouraged Gilgamesh to proceed, for it confirmed that he in fact had reached the "Landing Place."

In the morning the comrades attempted to penetrate the forest, careful to avoid "weapon-trees that kill." Enkidu found the gate, of which he had spoken to Gilgamesh. But as he tried to open it, he was thrown back by an unseen force. For twelve days he lay paralyzed.

When he was able to move and speak again, he pleaded with Gilgamesh: "Let us not go down into the heart of the forest." But Gilgamesh had good news for his comrade: while the latter was recovering from the shock, he—Gilgamesh—had found a tunnel. By the sounds heard from it, Gilgamesh was sure that it was connected to "the enclosure from which words of command are issued." Come on, he urged Enkidu; "do not stand by, my friend; let us go down together!"

Gilgamesh must have been right, for the Sumerian text states that

> Pressing on into the forest,
> the secret abode of the *Anunnaki*
> he opened up.

The entrance to the tunnel was grown over with (or hidden by) trees and bushes and blocked by soil and rocks. "While Gilgamesh cut down the trees, Enkidu dug up" the soil and rocks. But just as they made enough of a clearance, terror struck: "Huwawa heard the noise, and became angry." Now he appeared on the scene looking for the intruders. His appearance was "Mighty, his teeth as the teeth of a dragon; his face the face of a lion; his coming like the onrushing floodwaters." Most fearsome was his "radiant beam." Emanating from his forehead, "it devoured trees and bushes." From its killing force, "none could escape." A Sumerian cylinder seal depicted a god, Gilgamesh and Enkidu flanking a mechanical robot, no doubt the epic's "Monster with the Killing Beams" (Fig. 65).

It appears from the fragmented texts that Huwawa could armor himself with "seven cloaks," but when he arrived on the scene "only one he had donned, six are still off." Seeing this as their opportunity, the two comrades attempted to ambush Huwawa. As the monster turned to face his attackers, the Killing Beam from his forehead traced a path of destruction.

In the nick of time, rescue appeared from the heavens. Seeing their predicament, "down from the skies spoke to them divine Shamash." Do not try to escape, he advised them; instead, "draw near Huwawa." Then Shamash raised a host of swirling winds, "which beat against the eyes of Huwawa" and neutralized his beam. As Shamash had intended, "the radiant beams vanished, the brilliance became clouded." Soon, Huwawa was immobilized: "he is unable to move forward, nor is he able to move back." The two then attacked Huwawa: "Enkidu struck the guardian, Huwawa, to the ground. For two leagues the cedars resounded," so immense was the monster's fall. Then Enkidu "Huwawa put to death."

Exhilarated by their victory but exhausted by the battle, the two stopped to rest by a stream. Gilgamesh undressed to wash himself. "He cast off his soiled things, put on his clean ones; wrapped a fringed cloak about him, fastened with a sash." There was no need to rush: the way to the "secret abode of the Anunnaki" was no longer blocked.

Little did he know that a female's lust would soon undo his victory. . . .

The place, as stated earlier in the epic, was the "Crossroads of Ishtar." The goddess herself was wont to come and go from this "Landing Place." She too, like Shamash, must have watched the battle—perhaps from her aerial ("winged") Sky Chamber, as depicted on a Hittite seal (Fig. 66). Now, having seen Gilgamesh undress and bathe, "glorious Ishtar raised an eye at the beauty of Gilgamesh."

Approaching the hero, she minced no words about what was on her mind:

> Come, Gilgamesh, be thou my lover!
> Grant me the fruit of thy love.
> You be my man,
> I shall be your woman!

Fig. 65

Fig. 66

Promising him golden chariots, a magnificent palace, lord-ship over other kings and princes, Ishtar was sure she had enticed Gilgamesh. But answering her, he pointed out that he had nothing he could give her, a goddess, in return. And as to her "love," how long would that last? Sooner or later, he said, she would rid herself of him as of "a shoe which pinches the foot of its owner." Calling off the names of other men with whom she had been promiscuous, he turned her down. Enraged by this insulting refusal, Ishtar asked Anu to let the "Bull of Heaven" smite Gilgamesh.

Attacked by the Sky Monster, Gilgamesh and Enkidu forgot all about their mission, and ran for their lives. Aiding their escape back to Uruk, Shamash enabled them "the distance of a month and fifteen days, in three days to traverse." But on the outskirt of Uruk, on the Euphrates River, the Bull of Heaven caught up with them. Gilgamesh managed to reach the city, to summon its warriors. Outside the city walls, Enkidu alone remained to hold off the Sky Monster. When the Bull of Heaven "snorted," pits were opened in the earth, large enough to hold two hundred men each. As Enkidu fell into one of the pits, the Bull of Heaven turned around. Quickly Enkidu climbed out, and put the monster to death.

What exactly the Bull of Heaven was, is not clear. The Sumerian term—GUD.AN.NA—could also mean "Anu's attacker," his "cruise missile." Ancient artists, fascinated by the episode, frequently depicted Gilgamesh or Enkidu fighting with an actual bull, with the naked Ishtar (and sometimes Adad) looking on (Fig. 67a). But from the Epic's text it is clear that this weapon of Anu was a mechanical contraption made of metal and equipped with two piercers (the "horns") which were "cast from thirty minas of lapis, the coating on each being two fingers thick." Some ancient depictions show such a mechanical "bull," sweeping down from the skies (Fig. 67b).

After the Bull of Heaven was defeated, Gilgamesh "called out to the craftsmen, the armorers, all of them" to view the mechanical monster and take it apart. Then, triumphant, he and Enkidu went to pay homage to Shamash.

But "Ishtar, in her abode, set up a wail."

In the palace, Gilgamesh and Enkidu were resting from nightlong celebrations. But at the Abode of the Gods, the su-

a

b

Fig. 67

preme gods were considering Ishtar's complaint. "And Anu said to Enlil: 'Because the Bull of Heaven they have slain, and Huwawa they have slain, the two of them must die.' But Enlil said: 'Enkidu shall die, let Gilgamesh not die.'" Then Shamash interceded: it was done with his concurrence; why then should "innocent Enkidu die?"

While the gods deliberated, Enkidu was afflicted with a coma. Distraught and worried, Gilgamesh "paced back and forth before the couch" on which Enkidu lay motionless. Bitter tears flowed down his cheeks. As sorry as he was for his comrade, his thoughts turned to his own permeating anxiety: will he too lie one day dying like Enkidu? Will he, after all the endeavors, end up dead as a mortal?

In their assembly, the gods reached a compromise. The death sentence of Enkidu was commuted to hard labor in the depths of the mines—there to spend the rest of his days. To carry out the sentence and take him to his new home, Enkidu was told, two emissaries "clothed like birds, with wings for garments" shall appear unto him. One of them, "a young man whose face is dark, who like a Bird-Man is his face," shall transport him to the Land of the Mines:

> He will be dressed like an Eagle;
> By the arm he will lead thee.
> "Follow me," (he will say); he will lead you
> To the House of Darkness,
>   the abode below the ground;
> The abode which none leave who have entered into it.
> A road from which there is no return;
> A House whose dwellers are bereft of light,
>   where dust is in their mouths
>   and clay is their food.

An ancient depiction on a cylinder seal illustrated the scene, showing a Winged Emissary ("angel") leading Enkidu by the arm (Fig. 68).

Hearing the sentence passed on his comrade, Gilgamesh had an idea. Not far from the Land of Mines, he had learned, was the *Land of the Living:* the place whereto the gods had taken those humans who were granted eternal youth!

It was "the abode of the forefathers who by the great gods with the Purifying Waters were anointed." There, partaking of the food and beverage of the gods, have been residing

> Princes born to the crown
>   who had ruled the land in days of yore;
> Like Anu and Enlil, spiced meats they are served.
> From waterskins, cool water to them is poured.

Was it not the place whereto the hero of the Deluge, Ziusudra/Utnapishtim, had been taken—the very place from which Etana "to heaven ascended"?

Fig. 68

And so it was, that "the lord Gilgamesh, toward the Land of the Living set his mind." Announcing to the revived Enkidu that he would accompany him at least on part of his journey, Gilgamesh explained:

> O Enkidu,
> Even the mighty wither, meet the fated end.
> (Therefore) the Land I would enter,
> I would set up my *Shem.*
> In the place where the *Shems* have been raised up,
> I a *Shem* I would raise up.

However, proceeding from the Land of Mines to the Land of the Living was not a matter for a mortal to decide. In the strongest possible words, Gilgamesh was advised by the elders of Uruk and his goddess mother to first obtain the permission of Utu/Shamash:

> If the Land thou wish to enter,
>   inform Utu, inform Utu, the hero Utu!
> The Land, it is in Utu's charge;
> The Land which with the cedars is aligned,
>   it is the hero Utu's charge.
> Inform Utu!

Thus forewarned and advised, Gilgamesh offered a sacrifice to Utu, and appealed for his consent and protection:

> O Utu,
> The Land I wish to enter;
>   be thou my ally!
> The Land which with the cool cedars is aligned
> I wish to enter; be thou my ally!
> In the places where the *Shems* have been raised up,
> Let me set up my *Shem!*

At first, Utu/Shamash doubted whether Gilgamesh could qualify to enter the land. Then, yielding to more pleading and prayers, he warned him that his journey would be through a desolate and arid area: "the dust of the crossroads shall be thy dwelling place, the desert shall be thy bed . . . thorn and bramble shall skin thy feet . . . thirst shall smite thy cheeks." Unable to dissuade Gilgamesh, he told him that the "place where the *Shems* have been raised" is surrounded by seven mountains, and the passes guarded by fearsome "Mighty Ones" who can unleash "a scorching fire" or "a lightning which cannot be turned back." But in the end, Utu gave in: "the tears of Gilgamesh he accepted as an offering; like one of mercy, he showed him mercy."

But "the lord Gilgamesh acted frivolously." Rather than take the harsh overland road, he planned to cover most of the route by a comfortable sea voyage; after landing at the distant destination, Enkidu would go to the Land of Mines, and he (Gilgamesh) would proceed to the Land of the Living. He selected fifty young, unattached men to accompany him and Enkidu, and be rowers of the boat. Their first task was to cut and haul back to Uruk special woods, from which the MA.GAN boat—a "ship of Egypt"—was built. The smiths of Uruk fashioned strong weapons. Then, when all was ready, they sailed away.

They sailed, by all accounts, down the Persian Gulf, planning no doubt to circumnavigate the Arabian peninsula and then sail up the Red Sea toward Egypt. But the wrath of Enlil was swift to come. Had not Enkidu been told that a young "angel" would take him by the arm and bring him to the Land of Mines? How come, then, he was sailing with the joyful Gilgamesh, with fifty armed men, in a royal ship?

At dusk, Utu—who may have seen them off with great misgivings—"with lifted head went away." The mountains along the distant coast "became dark, shadows spread over them." Then, "standing alongside the mountain," there was someone who—like Huwawa—could emit rays "from which none can escape." "Like a bull he stood on the great Earth house"—a watchtower, it seems. The fearsome watchman must have challenged the ship and its passengers, for fear overcame Enkidu. Let us turn back to Uruk, he pleaded. But Gilgamesh would not hear of it. Instead, he directed the ship toward the shore, determined to fight the watchman—"that 'man,' if a man he be, or if a god he be."

It was then that calamity struck. The "three ply cloth"—the sail—tore apart. As if by an unseen hand, the boat capsized; and all in it sank down. Somehow, Gilgamesh managed to swim ashore; so did Enkidu. Back in the waters, they saw the sunken ship with its crew still at their posts, looking amazingly alive in their deaths:

After it had sunk, in the sea had sunk,
On the eve when the *Magan*-boat had sunk,
After the boat, destined to *Magan,* had sunk—
Inside it, as though still living creatures,
    were seated those who of a womb were born.

They spent the night on the unknown shore, arguing which way to go. Gilgamesh was still determined to reach "the land." Enkidu advised seeking a way back to "the city," Uruk. Soon, however, weakness overcame Enkidu. With passionate comradeship, Gilgamesh exhorted Enkidu to hold on to life: "My little weak friend," he fondly called him; "to the land I will bring thee," he promised him. But "Death, which knows no distinction," could not be held off.

For seven days and seven nights Gilgamesh mourned Enkidu, "until a worm fell out of his nose." At first he wandered aimlessly: "For his friend, Enkidu, Gilgamesh weeps bitterly as he ranges over the wilderness . . . with woe in his belly, fearing death, he roamed the wilderness." Again he was preoccupied with his own fate—"fearing death"—wondering: "When I die, shall I not be like Enkidu?"

Then his determination to ward off his fate took hold of him again. "Must I lay my head inside the earth, and sleep through all the years?" he demanded to know of Shamash. "Let mine eyes behold the sun, let me have my fill of light!" he begged of the god. Setting his course by the rising and setting Sun, "To the Wild Cow, to Utnapishtim the son of Ubar-Tutu, he took the road." He trod unbeaten paths, encountering no man, hunting for food. "What mountains he had climbed, what streams he had crossed—no man can know," the ancient scribes sadly noted.

At long last, as versions found at Nineveh and at Hittite sites relate, he neared habitations. He was coming to a region dedicated to *Sin*, the father of Shamash. "When he arrived at night at a mountain pass, Gilgamesh saw lions and grew afraid:"

He lifted his head to Sin and prayed:
"To the place where the gods rejuvenate,
    my steps are directed . . .
Preserve thou me!"

"As at night he lay, he awoke from a dream" which he interpreted as an omen from Sin, that he would "rejoice in Life." Encouraged, Gilgamesh "like an arrow descended amoung the lions." His battle with the lions has been commemorated pictorially not only in Mesopotamia, but throughout the ancient lands, even in Egypt (Fig. 69a, b, c).

After daybreak, Gilgamesh traversed a mountain pass. In the distance below, he saw a body of water, like a vast lake, "driven by long winds." In the plain adjoining the inland sea he could see a city "closed-up about"—a city surrounded by a wall. There, "the temple to Sin was dedicated."

a

b

c

Fig. 69

Outside the city, "close by the low-lying sea," Gilgamesh saw an inn. As he approached, he saw the "Ale-woman, Siduri." She was holding "a jug (of ale), a bowl of golden porridge." But as she saw Gilgamesh, she was frightened by his appearance: "He is clad in skins . . . his belly is shrunk . . . his face is like a wayfarer from afar." Understandably, "as the ale-woman saw him, she locked the door, she barred the gate." With great effort, Gilgamesh convinced her of his true identity and good intentions, telling her of his adventures and quest.

After Siduri let him rest, eat and drink, Gilgamesh was eager to continue. What is the best way to the Land of Living? he asked Siduri. Must he circle the sea and wind his way through the desolate mountains—or could he take a shortcut across the body of water?

> Now ale-woman, which is the way . . .
> What are its markers?
> Give me, O give me its markers!
> Suitably, by the sea I will go across;
> Otherwise, by the wilderness my course will be.

The choice, it turned out, was not that simple; for the sea he saw was the "Sea of Death":

> The ale-woman said to him, to Gilgamesh:
> "The sea, Gilgamesh, it is impossible to cross
> From days of long ago,
>   no one arrived from across the sea.
> Valiant Shamash did cross the sea,
>   but other than Shamash, who can cross it?
> Toilsome is the crossing,
>   desolate is its way;
> Barren are the Waters of Death
>   which it encloses
> How then, Gilgamesh, wouldst thou cross the sea?

As Gilgamesh remained silent, Siduri spoke up again, revealing to him that there might be, after all, a way to cross
the Sea of the Waters of Death:

> Gilgamesh,
> There is *Urshanabi,* boatman of Utnapishtim.
> With him are things that float,
>   in the woods he picks the things that bind together.
> Go, let he thy face behold.
> If it be suitable, with thee he shall cross;
> If it be not suitable, draw thou back.

Following her directions, Gilgamesh found Urshanabi the
boatman. After much questioning as to who he was, how
he had come hither, and where he was going, he was found
worthy of the boatman's services. Using long poles, they
moved the raft forward. In three days, "a run of a month and
fifteen days"—a forty-five day journey overland—"they left
behind."

He arrived at TIL.MUN—"The Land of the Living."

Whereto shall he go now? Gilgamesh wondered. You have
to reach a mountain, Urshanabi answered; "the name of the
mountain is *Mashu.*"

The instructions given by Urshanabi are available to us
from the Hittite version of the Epic, fragments of which
were found in Boghazkoy and other Hittite sites. From those
fragments (as put together by Johannes Friedrich: *Die hethi-
tischen Bruchstükes des Gilgamesh-Epos),* we learn that Gilgamesh was told to reach and follow "a regular way" which
leads toward "the Great Sea, which is far away." He was to
look for two stone columns or "markers" which, Urshanabi
vouched, "to the destination always bring me." There he had
to turn and reach a town named *Itla,* sacred to the god whom
the Hittites called *Ullu-Yah* ("He of the Peaks"?). He had to
obtain that god's blessing before he could go farther.

Following the directions, Gilgamesh did arrive at Itla. In the distance, the Great Sea could apparently be seen. There, Gilgamesh ate and drank, washed and made himself once again presentable as befits a king. There, Shamash once again came to his aid, advising him to make offerings to Ulluyah. Taking Gilgamesh before the Great God (Fig. 70), he urged Ulluyah: Accept his offerings, "grant him life." But Kumarbi, another god well known from Hittite tales, strongly objected: Immortality cannot be granted to Gilgamesh, he said.

Fig. 70

Realizing, it appears, that he would not be granted a *Shem,* Gilgamesh settled for second-best: Could he, at least, meet his forefather Utnapishtim? As the gods delayed their decision, Gilgamesh (with the connivance of Shamash?) left town and started to advance toward Mount *Mashu,* stopping each day to offer sacrifices to Ulluyah. After six days, he came unto the Mount; it was indeed the Place of the *Shems:*

> The name of the Mountain is *Mashu.*
> At the mountain of *Mashu* he arrived;
> Where daily the *Shems* he watched
> As they depart and come in.

The Mount's functions required it to be connected both to the distant heavens and to the far reaches of Earth:

> On high, to the Celestial Band
>> it is connected;
> Below,
>> to the Lower World it is bound.

There was a way to go inside the Mount; but the entrance, the "gate," was closely guarded:

> Rocket-men guard its gate.
> Their terror is awesome, their glance is death.
> Their dreaded spotlight sweeps the mountains.
> They watch over Shamash
>> as he ascends and descends.

(Depictions have been found showing winged beings or divine bull-men operating a circular beaming device mounted on a post; they could well be ancient illustrations of the "dreaded spotlight that sweeps the mountains"— Fig. 71a, b, c.)

"When Gilgamesh beheld the terrible glowing, his face

a

b

c

Fig. 71

he shielded; regaining his composure, he approached them."
When the Rocketman saw that the dreaded ray affected Gilgamesh only momentarily, he shouted to his partner: "He
who comes, of the flesh of the gods is his body!" The rays,
it appears, could stun or kill humans—but were harmless to
the gods.

Allowed to approach, Gilgamesh was asked to identify
himself and account for his presence in the restricted area.
Describing his partly divine origins, he explained that he had
come "in search of Life." He wished, he said, to meet his
forefather Utnapishtim:

> On account of Utnapishtim, my forefather,
>   have I come—
> He who the congregation of the gods had joined.
> About Death and Life I wish to ask him.

"Never was this achieved by a mortal," the two guards said.
Undaunted, Gilgamesh invoked Shamash and explained that
he was two-thirds god. What happened next is unknown, due
to breaks in the tablet; but at last the Rocketmen informed
Gilgamesh that permission was granted: "The gate of the
Mount is open to thee!"

(The "Gateway to Heaven" was a frequent motif on Near
Eastern cylinder seals, depicting it as a winged, ladder-
like gateway leading to the Tree of Life. It was sometimes
guarded by Serpents—Fig. 72).

Gilgamesh went in, following the "path taken by
Shamash." His journey lasted twelve *beru* (double-hours);
through most of it "he could see nothing ahead or behind";
perhaps he was blindfolded, for the text stresses that "*for
him,* light there was none." In the eighth double-hour, he
screamed with fear; in the ninth, "he felt a north wind fan-
ning his face." "When eleven *beru* he attained, dawn was
breaking." Finally, at the end of the twelfth double-hour, "in
brightness he resided."

He could see again, and what he saw was astounding. He

Fig. 72

saw "an enclosure as of the gods," wherein there "grew" a garden made up entirely of precious stones! The magnificence of the place comes through the mutilated ancient lines:

> As its fruit it carries carnelians,
>   its vines too beautiful to behold.
> The foliage is of lapis-lazuli;
> And grapes, too lush to look at,
>   of . . . stones are made . . .
> Its . . . of white stones . . .
> In its waters, pure reeds . . . of *sasu*-stones;
> Like a Tree of Life and a Tree of . . .
>   that of *An-Gug* stones are made . . .

On and on the description went. Thrilled and amazed, Gilgamesh walked about the garden. He was clearly in a *simulated* "Garden of Eden!"

What happened next is still unknown, for an entire column of the ninth tablet is too mutilated to be legible. Either in the artificial garden, or somewhere else, Gilgamesh finally encountered Utnapishtim. His first reaction on seeing a man from "days of yore" was to observe how much they looked alike:

> Gilgamesh said to him,
>   to Utnapishtim "The Far-away":
> "As I look upon thee, Utnapishtim,
> Thou are not different at all;
>   even as I art thou . . ."

Then Gilgamesh came straight to the point:

> Tell me,
> How joinest thou the congregation of the gods
>   in thy quest for Life?

In answer to this question, Utnapishtim said to Gilgamesh: "I will reveal to thee, Gilgamesh, a hidden matter; a secret of the gods I will tell thee." The secret was the *Tale of the Deluge:* How when he, Utnapishtim, was the ruler of Shuruppak and the gods resolved to let the Deluge annihilate Mankind, Enki secretly instructed him to build a special submersible vessel, and take aboard his family "and the seed of all living things." A navigator provided by Enki directed the vessel to Mount Ararat. As the waters began to subside, he left the vessel to offer sacrifices. The gods and goddesses—who circled Earth in their spacecraft while it was inundated—also landed on Mount Ararat, savoring the roasting meat. Finally, Enlil too landed, and broke into a rage when he realized that in spite of the oath taken by all the gods, Enki enabled Mankind to survive.

But when his anger subsided, Enlil saw the merit of such survival; it was then, Utnapishtim continued to recount, that Enlil granted him everlasting life:

> Thereupon, Enlil went aboard the ship.
> Holding me by the hand, he took me aboard.
> He took my wife aboard,
>   and made her kneel by my side.
> Standing between us,
>   he touched our foreheads to bless us:
> "Hitherto, Utnapishtim has been human;
> Henceforth, Utnapishtim and his wife
>   like gods shall be unto us.
> Far away shall the man Utnapishtim reside,
>   at the mouth of the water-streams."

And so it came to pass, Utnapishtim concluded, that he was taken to the Faraway Abode, to live among the gods. But how could this be achieved for Gilgamesh? "But now, who will for thy sake call the gods to Assembly, that the Life which thou seekest thou mayest find?"

On hearing the tale, and realizing that it is only the gods, in assembly, who can decree eternal life and that he, on his own, could not attain it—Gilgamesh fainted. For six days and seven nights he was totally knocked out. Sarcastically, Utnapishtim said to his wife: "Behold this hero who seeks Life; from mere sleep as mist he dissolves!" Throughout his sleep, they attended to Gilgamesh, to keep him alive, "that he may return safe on the way by which he came, that through the gate by which he entered he may return to his land."

Urshanabi the boatman was called to take Gilgamesh back. But at the last moment, when Gilgamesh was ready to leave, Utnapishtim disclosed to Gilgamesh yet another secret. Though he could not avoid death, he told him, there was a way to postpone it. He could do this by obtaining the secret plant which the gods themselves eat, to keep *Forever Young!*

> Utnapishtim said to him, to Gilgamesh:
> "Thou hast come hither, toiling and straining.
> What shall I give thee,
>   that thou mayest return to thy land?
> I will disclose, O Gilgamesh, a hidden thing;
> A secret of the gods I will tell thee:
> A plant there is,
>   like a prickly berrybush is its root.
> Its thorns are like a brier vine's,
>   thine hands they will prick.
> If thine hands obtain the plant,
> New Life thou wilt find."

The plant, we learn from what followed, grew underwater:

> No sooner had Gilgamesh heard this,
>   than he opened the water-pipe.
> He tied heavy stones to his feet;
> They pulled him down into the deep;
> He saw then the plant.
> He took the plant, though it pricked his hands.
> He cut the heavy stones from his feet;
> The second cast him back where he came from.

Going back with Urshanabi, Gilgamesh triumphantly said to him:

> Urshanabi,
> This plant is of all plants unique:
> By it a man can regain his full vigor!
> I will take it to ramparted Uruk,
>   there the plant to cut and eat.

Let its name be called
"Man Becomes Young in Old Age!"
Of this plant I shall eat,
        and to my youthful state shall I return.

A Sumerian cylinder seal, from circa 1700 B.C., which illustrated scenes from the epic tale, shows (at left) a half-
naked and unkempt Gilgamesh battling the two lions; on the
right, Gilgamesh holds up to Urshanabi the plant of everlasting youth. A god, in the center, holds an unusual spiral tool
or weapon (Fig. 73).

But Fate, as with all those who in the millenia and centuries that followed went in the search of the Plant of Youth,
intervened.

As Gilgamesh and Urshanabi "prepared for the night, Gilgamesh saw a well whose water was cool. He went down
to it to bathe in the water." Then calamity struck: "A snake
sniffed the fragrance of the plant. It came and carried off the
plant. . . ."

Thereupon Gilgamesh sits down and weeps
        his tears running down his face.
He took the hand of Urshanabi, the boatman.
"For whom," (he asked) "have my hands toiled?
For whom is spent the blood of my heart?
For myself, I have not obtained the boon;
        for a serpent a boon I affected. . . ."

Yet another Sumerian seal illustrates the epic's tragic end:
the winged gateway in the background, the boat navigated
by Urshanabi, and Gilgamesh struggling with the serpent.
Not having found Immortality, he is now pursued by the
Angel of Death (Fig. 74).

And so it was, that for generations thereafter, scribes copied and translated, poets recited, and storytellers related, the

Fig. 73

Fig. 74

tale of the first futile Search for Immortality, the epic tale of Gilgamesh.

This is how it began:

> Let me make known to the country
> Him who the Tunnel has seen;
> Of him who knows the seas,
>   let me the full story tell.
> He has visited the . . . (?) as well,
> The hidden from wisdom, all things . . .
> Secret things he has seen,
>   what is hidden from man he found out.
> He even brought tidings
>   of the time before the Deluge.
> He also took the distant journey,
>   wearisome and under difficulties.
> He returned, and upon a stone column
>   all his toil he engraved.

And this, according to the Sumerian King Lists, is how it all ended:

> The divine Gilgamesh, whose father was a human, a high priest of the temple precinct, ruled 126 years. Ur-lugal, son of Gilgamesh, ruled after him.

VIII

Riders of the Clouds

The journey of Gilgamesh in search of Immortality has undoubtedly been the fountainhead of the many tales, in subsequent millenia, of demi-gods or heroes claiming such a status, who have likewise gone to search for paradise on Earth or to reach the Celestial Abode of the Gods. Without question, the detailed Epic of Gilgamesh also served as a guide book in which the subsequent searchers sought to find the ancient landmarks by which the Land of the Living could be reached and the way to it ascertained.

The similarities between the geographic landmarks; the man-made (or rather the god-made) tunnels, corridors, air locks and radiation chambers; and the bird-like beings, or "Eagles," as well as many other major and minor details—are far too numerous and identical to be mere accidents. At the same time, the epic tale of the journey can explain the confusion that reigned millenia later concerning the exact location of the cherished target; because as our detailed analysis has shown, Gilgamesh made not one but two journeys—a fact generally ignored by modern scholars and possibly also by past ones.

The Gilgamesh drama reached its culmination in the Land of *Tilmun,* an Abode of the Gods and a place of the *Shems.* It was there that he encountered an ancestor who had escaped mortality, and had found the secret plant of eternal youth. It was there that other divine encounters, as well as events affecting the course of human history, occurred in later millenia. It was there, we believe, that the *Duat* was—the Stairway to Heaven.

But that was not the first destination of Gilgamesh, and we ought to follow in his footsteps in the same sequence by which he himself had embarked on his journey: his first destination on the road to Immortality was not Tilmun, but the "Landing Place" on the Cedar Mountain, within the great Cedar Forest.

Scholars (e.g. S. N. Kramer, *The Sumerians*) have termed as "cryptic and still enigmatic" Sumerian statements that Shamash could "rise" in the "Cedar Land," and not only in Tilmun. The answer is that apart from the Spaceport at Tilmun, from which the farthest heavens could be reached, there was also a "Landing Place" from which the gods "could scale the skies" of Earth. This realization is supported by our conclusion, that the gods indeed had two types of craft: the GIRs, the Rocketships that were operated from Tilmun; and what the Sumerians called a MU, a "Sky Chamber." It is a credit to the technology of the Nefilim that the uppermost section of the GIR, the Command Module—what the Egyptians called *Ben-Ben*—could be detached and fly in Earth's skies as a MU.

The ancient peoples had seen the GIRs in their silos (Fig. 27) or even in flight (Fig. 75). But they depicted more frequently the "Sky Chambers"—vehicles which we would nowadays classify as UFOs (Unidentified Flying Objects). The one the patriarch Jacob had seen in his vision might well have looked like the Sky Chamber of Ishtar (Fig. 66); the Flying Wheel described by the prophet Ezekiel was akin to the Assyrian depictions of their Flying God roaming the skies, at cloud level, within a spherical Sky Chamber (Fig. 76a). Depictions found at an ancient site across the Jordan from Jericho, suggest that for landing these spherical vehicles extended three legs (Fig. 76b); they could well have been the fiery Whirlwinds in which the prophet Elijah was carried off heavenwards at that very same location.

As the Sumerian "Eagles," so were the Flying Gods of antiquity depicted by all ancient peoples as gods equipped with wings—the Winged Beings to whose depictions we can trace the Judeo-Christian acceptance of the winged *Cherubim* and Angels (literally: Emissaries) of the Lord (Fig. 77).

Tilmun, then, was the location of the Spaceport. The Cedar Mountain was the location of the "Landing Place," the "Crossroads of Ishtar,"—the Airport of the Gods. And it was to the latter that Gilgamesh had first set his course.

While the identification of Tilmun and its location is no mean challenge, there is little problem in locating the Cedar Forest. With the exception of subsidiary growths on the island of Cyprus, there is only one such location throughout the Near East: the mountains of Lebanon. These majestic cedar trees, which can reach 150 feet in height, were repeatedly extolled in the Bible and their uniqueness was

Fig. 75

a

b

Fig. 76

Fig. 77

known to the ancient peoples from the earliest times. As the biblical and other Near Eastern texts attest, the Cedars of Lebanon were earmarked for the construction and decoration of temples ("gods' houses")—a practice described in detail in I Kings, in the chapters dealing with the building of the Jerusalem Temple by Solomon (after the Lord Yahweh had complained "Why build ye not me a House of cedar?").

The biblical Lord appears to have been quite familiar with the cedars, and frequently employed them in his allegories, comparing rulers or nations to cedars: "Assyria was a cedar in Lebanon, with fair branches and a shadowing shroud and of high stature . . . waters nourished it, subterranean streams gave it height"—until the wrath of Yahweh toppled it and smashed its branches. Man, it appears, had never been able to cultivate these cedars; and the Bible records an attempt that had completely failed. Attributing the attempt to the king of Babylon (factually or allegorically), it is stated that "He came to Lebanon and took the cedar's highest branch," selecting off it a choice seed. This seed "he planted in a fruitful field, he placed it by great waters." But what grew up was not a tall cedar—only a willow-like tree, "a spreading vine of low stature."

The biblical Lord, on the other hand, knew the secret of cedar cultivation:

> Thus sayeth the Lord Yahweh:
> "From the cedar's crest, from its topmost branches a tender shoot I will take; and I will plant it upon a high and steep mountain . . .
> And it will put forth branches, and bear fruit, and become a mighty cedar."

This knowledge apparently stemmed from the fact that the cedar grew in the "Orchard of the gods." There, no other tree could match it; "it was the envy of all the trees that were in Eden, the garden of the gods." The Hebrew term *Gan* (orchard, garden), stemming as it does from the root *gnn* (protect, guard), conveys the sense of a guarded and restricted area—the same sense as is imparted to the reader of the Gilgamesh narrative: a forest that extends "for many leagues," watched over by a Fiery Warrior ("a terror to mortals"), accessible only through a gateway which paralyzed the intruder who touched it. Inside, there was "the secret abode of the Anunnaki"; a tunnel led to "the enclosure from which

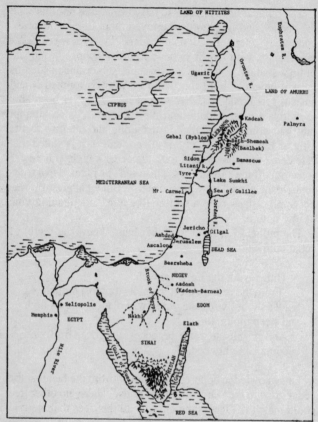

Fig. 78

words of command are issued"—"the underground place of Shamash."

Gilgamesh almost made it to the Landing Place, for he had the permission and help of Shamash. But the wrath of Ishtar (when he turned down her advances) completely reversed the course of events. Not so, according to the Old

Testament, was the fate of another mortal king. He was the king of Tyre—a city-state on the coast of Lebanon, a short distance away from the cedar mountains; and the Deity (as told in chapter 28 of the Book of Ezekiel) did enable him to visit the Sacred Mountain:

> Thou hast been in Eden, the Garden of God; every precious stone was thy thicket . . .
> Thou art an annointed Cherub, protected; and I have placed thee in the Sacred Mountain.
> As a god werest thou, moving within the Fiery Stones.

Gilgamesh sought to enter the Landing Place of the gods uninvited; the king of Tyre not only was permitted to come to the place, but evidently was also given a ride in "the fiery stones," flying as a Cherub. As a result, "a god I am," he said; "in the Abode of the Deity I sat, in the midst of the waters." For his haughtiness of heart, the prophet was to inform him, he was to die the death of a heathen by the hands of strangers.

Both the Hebrews of biblical times, and their neighbors to the north, were thus acquainted with the location and nature of the Landing Place in the Cedar Mountain which Gilgamesh attempted to penetrate in a previous millenium. It was, as we shall show, not a "mythological" place, but a real one: not only texts, but also pictorial depictions exist from those ancient days, attesting to the existence and functions of the place.

In the tale of the king who tried to grow a cedar, the Old Testament reports that he "carried off the twig to a land of commerce," and planted the seed "in a city of merchants." Such a land and such cities-of-merchants need not have been looked-for far: along the coast of Lebanon, from where Anatolia begins in the north to Palestine in the south, there were several Canaanite coastal cities whose wealth and power grew with their international commerce. Best known from

biblical narratives were Tyre and Sidon; centers of trade and
shipping for millenia, their fame reached its peak under their
Phoenician rulers.

In ruins and buried under a mount since its destruction by
Assyrian invaders lay yet another city, perhaps the northernmost outpost of the Canaanites at the borders of the Hittite
empire. Its remains were accidentally uncovered in 1928 by
a farmer who set out to plough a new field near the mount
called Ras Shamra. The extensive excavations that followed
uncovered the ancient city of *Ugarit.* The spectacular finds
included a large palace, a temple to the god *Ba'al* ("The
Lord"), and a variety of artifacts. But the real treasures were
scores of clay tablets inscribed in an alphabetic cuneiform
script (Fig. 79), written in a "Western-Semitic" language
akin to biblical Hebrew. The tablets, whose contents were
first presented by Charles Virollaud over many years in the
scientific periodical *Syria,* retrieved from relative obscurity
the Canaanites, their life and customs, and their gods.

At the head of the Canaanite pantheon was a supreme
deity called *El*—a term which in biblical Hebrew was the
generic term for "deity," stemming as it did from the Akkadian word *Ilu,* which literally meant "Lofty One." But in the
Canaanite tales of gods and men. *El* was the personal name
of an actual deity, who was the final authority in all affairs
be they divine or human. He was father of the gods, as well
as *Ab Adam* ("father of men"); his epithets were The Kindly,
The Merciful. He was "creator of things created" and the
"one who alone could bestow kingship." A stela found in
Palestine (Fig. 80) depicts El seated on his throne and being
served a beverage by a younger deity, probably one of his
many sons. El wears the conical, horned headdress which
was the recognition mark of the gods throughout the ancient
Near East; and the scene is dominated by the omnipresent
Winged Globe, emblem of the Planet of the Gods.

In "olden times," El was a principal deity of Heaven and
Earth. But at the time at which the events related in the tablets had taken place, El lived in semi-retirement, aloof from

Fig. 79

Fig. 80

daily affairs. His abode was "in the mountains," at "the two headwaters." There he sat in his pavilion, receiving emissaries, holding councils of the gods, and trying to resolve the recurring disputes among the younger gods. Many of these were his own children: some texts suggest that El may have had seventy offspring. Of them, thirty were by his official consort Asherah (Fig. 81); the others, by an assortment of concubines or even by human females. One poetic text tells how two females saw El naked as they were strolling on the beach; they were completely charmed by the size of his penis, and ended up each bearing him a son. (This attribute of El is prominent in a depiction of him, as a winged god, on a Phoenician coin—Fig. 82).

El's principal children, however, were three sons and one daughter: the gods *Yam* ("Ocean, Sea"), *Ba'al* ("Lord") and

Fig. 81

Fig. 82

Mot ("Smiter, Annihilator") and the goddess *Anat* ("She Who Responded"). In names and relationships they clearly paralleled the Greek gods Poseidon (God of the Seas), Zeus (Lord of the Gods) and Hades (God of the Lower World). Ba'al, as Zeus, was always armed with a lightning-missile (Fig. 82), the bull his cult symbol. When Zeus fought Typhon, it was his sister Athena, goddess of War and Love, who alone stood by him; and in Egyptian tales, Isis alone stood by her brother-husband Osiris. So it was when Ba'al fought his two brothers: his sister-lover Anat alone came to his help. Like Athena, she was on the one hand "The Maiden," often flaunting her naked beauty (Fig. 82); and on the other hand the Goddess of War, the lion a symbol of her bravery (Fig. 83). (The Old Testament called her *Ashtoreth*.)

The links to Egyptian prehistorical recollections and beliefs were no less obvious than to those of Greece. Osiris was resurrected by Isis after she had found his remains at the Canaanite city of Byblos. Likewise, Ba'al was brought back to life by Anat after he was smitten by Mot. Seth, the adversary of Osiris, was sometimes called in Egyptian writings "Seth of *Saphon*"; Ba'al, as we shall see, acquired the title "Lord of *Zaphon*." Egyptian monuments of the New Kingdom—paralleling the Canaanite period—often depicted the Canaanite gods as Egyptian deities, calling them Min, Reshef, Kadesh, Anthat (Fig. 84). We thus find the same tales applying to the same gods, but under different names, throughout the ancient world.

Scholars have pointed out that all these tales were echoes, if not actual versions, of the much earlier and original Sumerian tales: not only of Man's Search for Immortality, but also of love, death and resurrection among the gods. All along, the tales are replete with episodes, details, epithets, and teachings which also fill the Old Testament—attesting to a common locale (greater Canaan), common traditions and common original versions.

One such text is the tale of *Danel* (Dan-El—"El's judge"—*Daniel* in Hebrew), a righteous chieftain who could

Fig. 83

Fig. 84

not beget a rightful heir. He appealed to the gods to give him one, so that when he died, the son could erect a stela in his memory at *Kadesh*. From this we surmise that the area of the tale's events was where southern Canaan (the *Negev*) merges into the Sinai peninsula, for it was there that Kadesh ("The Sacred" city) was located.

Kadesh was encompassed in the territory of the biblical Patriarch Abraham; and the Canaanite tale of Danel is indeed replete with similarities to the biblical tale of the birth of Isaac to the aging Abraham and Sarah. Much as in the Book of Genesis, we read in the Canaanite tale that Danel, getting on in years without a male heir, saw his chance to get divine help when two gods arrived at his habitat. "Forthwith . . . he gives offering to the gods to eat, gives offering to drink to the Holy Ones." The divine guests—who turn out to be El, "The Dispenser of Healing," and Ba'al—stay with Danel a whole week, during which he overwhelms them with his supplications. Finally Ba'al "approaches El with his (Danel's) plea." Yielding, El "by the hand takes his servant" and grants him *"Spirit"* whereby Danel's virility is restored:

> With life-breath Danel is quickened . . .
> With life-breath he is invigorated.

To the disbelieving Danel, El promises a son. Mount your bed, he tells him, kiss your wife, embrace her . . . "by conception and pregnancy she will bear a male child to Danel." And just as in the biblical tale, the matriarch does bear a Rightful Heir, and the Succession is assured. They name him *Aqhat;* the gods nickname him *Na'aman* ("The Pleasant One").

As the boy grows to be a young man, the Craftsman of the Gods presents him with a gift of a unique bow. This soon arouses the envy of Anat, who wishes to possess the magical bow. To get it, she promises Aqhat anything he would like to have—silver, gold, even Immortality:

> Ask for Life, O Aqhat the youth—
> Ask for Life and I will grant it to thee;
> For Immortality (ask),
>   and I will bestow it upon thee.
> With Baal I will make thee count the years;
> With the sons of El shalt thou count the months.

Moreover—she promised—not only would he live as long as the gods, but he would be invited to join them for the Lifegiving ceremony:

> And Baal, when he grants Life,
>   a feast he gives;
> A banquet he holds for the One-Given-Life.
> He bids him a drink,
>   sings and chants over him sweetly.

But Aqhat does not believe that Man can escape his mortal fate, and does not wish to part with the bow:

> Lie not, O Maiden—
>   to a hero thy lies are loathesome.
> How can a mortal an Afterlife acquire?
> How can a mortal Eternity obtain? . . .
> The death of all men will I die;
> Yea, I shall surely die.

He also points out to Anat that the bow was made for warriors like himself and not for a female to use. Insulted, Anat "traverses the land" to El's abode, to seek permission to smite Aqhat. El's enigmatic response permits punishment only up to a point.

Now Anat turns to cunning. "Over a thousand fields, ten thousand acres" she travels back to Aqhat. Pretending to be at peace and in love with him, she laughs and giggles. Addressing him as "Aqhat the youth," she states: "Thou are my brother, I am thy sister." She persuades him to accompany her to the city of "The Father of the gods; the Lord of the Moon." There she asks *Taphan* to "slay Aqhat for his bow" but then "make him live again"—to put Aqhat to temporary death, only long enough for Anat to take away his bow. Taphan, following Anat's instructions, "smites Aqhat twice on the skull, thrice above the ear," and Aqhat's "soul escapes like vapor." But before Aqhat can be revived—if Anat had ever so intended—his body is ravaged by vultures. The terrible news is brought to Danel as, "sitting before the gate, under a mighty tree," he "judges the cause of the widow, adjudicates the case of the orphan." With Ba'al's aid, a search is instituted for the dismembered Aqhat, but to no avail. In revenge, Aqhat's sister, in disguise, travels to the abode of Taphan and, getting him drunk, attempts to slay him. (A possible happy end, wherein Aqhat was resurrected after all, is missing.)

The transfer of action from the mountains of Lebanon to the "City of Lord Moon" is also an element found in the Gilgamesh epic. Throughout the ancient Near East, the deity associated with the Moon was Sin (Nannar in the original Sumerian). His Ugaritic epithet was "Father of Gods"; he indeed was the father of Ishtar and her brother. The first attempt of Gilgamesh, to reach his goal via the Landing Place in the Cedar Mountain, was frustrated by Ishtar, who sought to have him killed by the Bull of Heaven after he rejected her advances. Undertaking the second journey toward the Land Tilmun, Gilgamesh too arrived at a walled city "whose temple to Sin was dedicated."

But whereas Gilgamesh arrived in the region of Sin after a long and hazardous trek, Anat—like Ishtar—could get around from place to place in no time—for she neither walked nor traveled on assback; instead, she flew from place to place. Many Mesopotamian texts referred to Ishtar's flying journeys and her ability to roam in Earth's skies, "crossing heaven, crossing earth." A depiction of her in her temple in Ashur, an Assyrian capital, showed her wearing goggles, a tight-fitting helmet and extended "earphones" or panels (see

Fig. 58). In the ruins of Mari on the Euphrates River, a life-sized statue of a goddess was found, equipped with a "black box," a hose, a horned helmet with built-in earphones and other attributes of an aeronaut (Fig. 85). This ability "as a bird to fly," also attributed to the Canaanite deities, features in all the epic tales discovered at Ugarit.

One such tale, in which a goddess flies to the rescue, is a text titled by scholars "The Legend of King Keret"—*Keret* being capable of interpretation as the king's personal name, or the name of his city ("The Capital"). The tale's main theme is the same as that of the Sumerian epic of Gilgamesh: Man's striving for Immortality. But it begins like the biblical tale of Job, and has other strong biblical similarities.

Job, according to the biblical tale, was a righteous and "pure" man of great wealth and power who lived in the "Land of *Utz*" (the "Land of Advice"), a land in the domain of the "Children of the East." All went well until "one day, when the sons of the gods came to present themselves to the Lord, Satan came also among them." Persuading the Lord to test Job, Satan was permitted to afflict him first with the loss of his children and all his wealth, and then with

Fig. 85

all manner of sickness. As Job sat mourning and suffering, three of his friends came to console him; the *Book of Job* was composed as a record of their discussions concerning matters of life and death and the mysteries of Heaven and Earth.

Bewailing the turn of events, Job longed for the days of yore, when he was honored and respected: "at the gates of *Keret*, in the public square, my seat was at the ready." In those days, Job reminisced, he believed that "as the Phoenix shall be my days, with my Establisher shall I die." But now, with nothing left and afflicted with illnesses, he felt like dying then and there.

The friend who had come from the south reminded him that "Man is born unto travail; only the son of *Reshef* can to the heights fly up": Man was mortal after all, so why this extreme agitation?

But Job answered enigmatically that it was not so simple: "The Lord's Essence is within me," he said; "its radiance feeds my *Spirit*." Was he disclosing, in the hitherto uncomprehended verse, that he was of partly divine blood? That, therefore, like Gilgamesh, he had expected to live as long as the ever-rejuvenating Phoenix, to die only when his "Establisher" shall die. But now he realized that "Eternally I shall no longer live; like vapor are my days."

The tale of Keret too depicts him at first as a prosperous man who loses in quick succession his wife and children through sickness and war. "He sees his offspring ruined . . . in its entirety a posterity perishing," and realizes that it is the end of his dynasty: "wholly undermined is his throne." His mourning and grief grow daily; "his bed is soaked by his weeping." Daily "he enters the inner chamber" in the temple and cries to his gods. Finally, El "descends unto him" to find out "what ails Keret that he weeps." It is then that the texts disclose that Keret is partly divine, for it was El who had fathered him (by a human female).

El advises his "beloved lad" to stop mourning, and to remarry, for he would be blessed with a new heir. He is told to wash and make himself presentable, and go seek the hand of the daughter of the king of Udum (possibly the biblical Edom). Keret, accompanied by troops and laden with gifts, goes to Udum and does as El had instructed. But the king of Udum turns down all the silver and gold. Knowing that Keret "is the flesh of the Father of Men"—of divine origins—he asks for a unique dowry: let the firstborn son that his daughter shall bear Keret also be semi-divine!

The decision is, of course, not up to Keret. El, who had given him the marriage counsel to begin with, is not available. Keret therefore sets his steps to the shrine of Asherah, and seeks her help. The next scene takes place at the abode of El, where the appeal of Asherah is supported by the younger gods:

> Then came the companies of the gods,
> And puissant Ba'al spake up:
> "Now come, O Kindly One, El benign:
> Will thou not bless Keret the pure blooded,
>   nor please the beloved lad of El?"

Thus prodded, El consents and "blesses Keret," promising him that he shall have seven sons and several daughters. The firstborn son, El announces, is to be named *Yassib* ("Permanent") for indeed he will be granted permanence. This will be achieved because when he is born, not his mother but rather the goddesses Asherah and Anat will suckle him. (The theme of a king's child being nursed by a goddess, thereby being granted longer life, was depicted in the art of all the Near Eastern peoples—Fig. 86).

The gods keep their promises; but Keret, growing in wealth and power, forgets his vows; in the manner of the king of Tyre in the prophecies of Ezekiel, his heart grew haughty, and he began to boast to his children about his divine origins. Angered, Asherah afflicted him with a fatal disease. As it became clear that Keret was on the verge of death, his sons were astonished: How can this happen to Keret, "a son of El,

Fig. 86

an offspring of the Kindly One, a holy being?" In disbelief, the sons question their father—for surely his failed claim to Immortality has a bearing on their own lives as well:

> In thy Life, father, we rejoiced;
> We exalted in thy Not-dying . . .
> Wilt thou die then, father, like the mortals?

Their father's silence speaks for itself, and now the sons turn to the gods:

> How can it be said,
> "A son of El is Keret,
>   an offspring of the Kindly One
>   and a holy being?"
> Shall then a god die?
> An offspring of the Kindly One not live?

Embarrassed, El asks the other gods: "Who among the gods can remove the sickness, drive out the malady?" Seven times El issued this appeal, but "none among the gods answers him." In desperation, El appeals to the Craftsman of the Gods and his assistants, the Crafts-goddesses who know all magic. Responding, the "female who removes illness," the goddess Shataqat, takes to the air. "She flies over a hundred towns, she flies over a multitude of villages. . . ." Arriving in the house of Keret in the nick of time, she manages to revive him.

(The tale, though, has no happy end. Since Keret's claim to Immortality had proven vain, his firstborn son suggested that Keret abdicate in his favor. . . .)

Of greater importance to the understanding of ancient events are the several epic tales dealing with the gods themselves. In these, the ability of the gods to fly about is accepted as a matter of course; and their haven in the "Crest of Zaphon" is featured as the aeronauts' resting place. The central figures in these tales are Ba'al and Anat, the brother-sister who are also lovers. Ba'al's frequent epithet is *"The Rider of the Clouds"*—an epithet which the Old Testament has claimed for the Hebrew deity. Anat's own flying capabilities, which became apparent in the tales dealing with the relations between gods and men, are even more highlighted in the tales of the gods themselves.

In one such text, Anat is told that Ba'al went to fish "in the meadow of *Samakh*" (Fig. 87). The area happens to still be called by that very name to this day: it is *Lake Sumkhi* ("Lake of Fishes") in northern Israel, where the Jordan River begins to flow into the Sea of Galilee; and it is still renowned for its fishes and wildlife. Anat decided to join Ba'al there:

> She raises wing, the Maiden Anat,
>> she raises wing and tours about flying
>> to the midst of the meadow of Samakh
>> which with buffaloes abounds.

Seeing her, Ba'al signaled her to come down; but Anat began to play hide-and-seek. Annoyed, Ba'al asked whether she expects him to "anoint her horns"—a lovemaking expression—"while in flight." Unable to find her, Ba'al took off "and went up . . . in the skies" unto his throne-seat on the "Crest of Zaphon." The playful Anat soon appeared there too, "upon Zaphon in pleasure (to be)."

Fig. 87

The idyllic get-together, however, could take place only in later years, when the position of Ba'al as the Prince of Earth and acknowledged master of the northern lands was finally established. Earlier, Ba'al engaged in life-and-death struggles with other contenders for the godly throne; the prize of all these fights was a place known as *Zarerath Zaphon*—commonly translated "the Heights of Zaphon," but specifically meaning "The Rocky Crest in the North."

These bloody struggles for dominion over certain strongholds or lands were compounded by the positioning for succession, as the head of the pantheon grew old and semi-retired. Conforming to marriage traditions first reported in the Sumerian writings, El's official consort Asherah ("the Ruler's daughter") was his half-sister. This made the first son born by her the rightful heir. But, as had happened before, he was often challenged by the firstborn—a son who chronologically was born first, but by another mother. (The fact that Ba'al, who had at least three wives, could not marry his beloved Anat confirms that she was a full, rather than a half-sister of his.)

The Canaanite tales begin at the remote mountainous

abode of El, where he secretly bestows the succession upon
Prince Yam. The goddess Shepesh, "Torch of the Gods,"
comes flying to Ba'al to reveal to him the bad news: "El is
overturning the Kingship!" she shouts in alarm.

Ba'al is advised to present himself to El, and put the dispute before the "Assembled Body"—Council—of the gods.
His sisters counsel him to be defiant:

> There now, be off on your way
>   toward the Assembled Body,
>   in the midst of Mount Lala.
> At the feet of El do not fall down,
>   prostrate thyself not unto the Assembled Body;
> Proudly standing, say ye your speech.

Learning of the ploy, Yam sends his own emissaries to
the gathered gods, to demand that the rebellious Ba'al be
surrendered to him. "The gods were sitting to eat, the Holy
Ones for dinner; Ba'al was attending upon El" when the
emissaries entered. In the hush that followed, they present
Yam's demand. To indicate that they meant business, "at
El's feet they do not fall down"; they hold their weapons
at the ready: "Eyes that are like a whetted sword, flashing a consuming fire." The gods drop to the ground and
take cover. El is willing to yield Ba'al. But Ba'al seizes his
own weapons and is about to jump the emissaries, when
his mother restrains him: an emissary bears immunity, she
reminds him.

As the emissaries return to Yam empty-handed, it is
clear that there is no way but for the two gods to meet on
the battlefield. A goddess—perhaps Anat—conspires with
the Craftsman of the Gods to provide Ba'al with two divine
weapons, the "chaser" and the "thrower" which "swoop like
an eagle." Meeting in combat, Ba'al overwhelms Yam and is
about to "smash Yam," when the voice of Asherah reaches
him: spare Yam! Yam is allowed to live, but is banished to
his maritime domains.

In return for sparing Yam, Ba'al asks Asherah to support his quest for supremacy over the Crest of Zaphon. Asherah is resting at a seaside resort, and it is quite reluctantly that she makes the journey to El's abode in a hot and dry place. Arriving "thirsty and parched," she puts the problem before him and asks that he decide with wisdom, not emotion: "Thou art great indeed and wise," she flatters him; "thy beard's gray hair instructs thee . . . Wisdom and Everlife are thy portion." Weighing the situation, El agrees: let Ba'al be the master of the Crest of Zaphon; let him build his house there.

What Ba'al has in mind, however, is not just an abode. His plans require the services of *Kothar-Hasis* ("The Skilled and Knowing"), Craftsman of the Gods. Not only modern scholars, but even the first century Philo of Byblos (quoting earlier Phoenician historians) have compared Kothar-Hasis with the Greek divine craftsman Hephaestus, who built the abode of Zeus and Hera. Others find parallels with the Egyptian Thoth, god of artcraft and magic. Indeed, the Ugaritic texts state that the emissaries sent to fetch Kothar-Hasis were to look for him in Crete and Egypt; presumably it was in those lands that his skills were being employed at that time.

When Kothar-Hasis arrived at Ba'al's place, the two went over the construction plans. It turned out that Ba'al desired a two-part structure, one an *E-khal* (a "great house") and the other a *Behmtam,* commonly translated "house" but which literally means *"a raised platform."* There was some disagreement between the two regarding where a funnel-like window, which could open and close in some unusual manner, should be placed. "Thou shalt heed my words, O Ba'al," Kothar-Hasis insisted. When the structure was completed, Ba'al was concerned that his wives and children would be hurt. To allay his fears, Kothar-Hasis ordered that trees of Lebanon, "from *Sirion* its precious cedars," be piled up within the structure—and started up a fire. For a full week the fire burnt intensely; silver and gold within it melted down; but the structure itself was neither damaged nor destroyed.

The underground silo and raised platform were ready!
  Losing no time, Ba'al decided to test the facility:

> Ba'al opened the Funnel in the Raised Platform,
>   the window within the Great House.
> In the clouds, Ba'al opened rifts.
> His godly sound Ba'al discharges. . . .
> His godly sound convulses the earth.
> The mountains quake. . . .
> A-tremble are the . . .
> In east and west, earth's mounts reel.

As Ba'al soared skyward, the divine messengers Gapan and Ugar joined him in flight: "The winged ones, the twain, flock the clouds" behind Ba'al; "bird-like the twain" soared above the snow-covered peaks of Zaphon. But with the new facilities, the Crest of Zaphon was turned into the "Fastness of Zaphon"; and Mount *Lebanon* ("The White One," after its snowy peaks) acquired the epithet *Sirion*—"The Armored" Mountain.

Attaining mastery over the Fastness of Zaphon, Ba'al also acquired the title *Ba'al Zaphon*. The title simply means "Lord of Zaphon," of the Northern Place. But the original connotation of the term *Zaphon* was not geographical; it meant both "the hidden-away," and "the observation place." Undoubtedly, all these connotations played a role in naming Ba'al "Lord of Zaphon."

Now that he had attained these powers and prerogatives, Ba'al's ambitions grew in scope. Inviting the "sons of the gods" to a banquet, he demanded fealty and vassalage; those who disagreed were set upon: "Ba'al seizes the sons of Ash-erah; Rabbim he strikes in the back, Dokyamm he strikes with a bludgeon." While some were slaughtered, others escaped. Drunk with power, Ba'al mocked them:

> Ba'al's enemies take to the woods;
> His enemies hide in the side of the mountain.
> Puissant Ba'al shouts:
> "O enemies of Ba'al why do you quake?
>   Why do you run, why do you hide?"
> The Eye of Ba'al splinters up;
> His stretched hand the cedar breaks;
> His right (hand) is mighty.

Pursuing his quest for mastery, Ba'al—with the aid of Anat—battled and annihilated such male adversaries as "Lothan, the serpent," Shalyat, "the seven-headed dragon," Atak "the Bullock," as well as the goddess Hashat, "the Bitch." We know from the Old Testament that Yahweh, the biblical Lord, was also a bitter adversary of Ba'al; and as Ba'al's influence grew among the Israelites when their king married a Canaanite princess, the prophet Elijah arranged a contest between Ba'al and Yahweh upon Mount Carmel. When Yahweh prevailed, the three hundred priests of Ba'al were promptly executed. In this adversity, it was for Yahweh that the Old Testament claimed mastery over the Crest of Zaphon. Significantly, the claims were made in almost identical language, as Psalm 29 and other verses make clear:

> Give unto Yahweh, O sons of the gods,
> Unto Yahweh give homage and supremacy.
> Give unto the Lord the homage of his *Shem;*
> Bow down unto Him in his Sacred Splendor.
> The sound of the Lord is upon the waters:
> The Lord of Glory thundereth,
>   echoes upon the many waters.
> His sound is powerful, full of majesty.
> The sound of the Lord the cedars breaks;

The cedars of Lebanon Yahweh splinters.
He makes *Lebanon* skip as a calf, and
Sirion like a young buffalo.
His sound from amidst the fiery flames cuts through. . . .
The Lord in his Great House is glorified.

As Ba'al in the Canaanite texts, so was the Hebrew Deity "Rider of the Clouds." The Prophet Isaiah envisioned Him flying south toward Egypt, "riding swiftly upon a cloud, he shall descend upon Egypt; the gods of Egypt shall quail before him." Isaiah also claimed to have personally seen the Lord and His winged attendants:

In the year that king Uzziah died, I beheld and saw the Lord seated upon a high and raised up Throne; its Lifters filled up the Great House. The Fire-Attendants were stationed above it, six wings, six wings to each of them . . . The threshold beams were shaken by the sound, and the House was filled with smoke.

The Hebrews were forbidden to worship, and therefore to make, statues or engraved images. But the Canaanites, who must have known of Yahweh, as the Hebrews had known of Ba'al, left us a depiction of Yahweh as conceived by them. A fourth century B.C. coin which bears the inscription *Yahu* ("Yahweh") depicts a bearded deity seated upon a throne shaped as a winged wheel (Fig. 88).

Fig. 88

It was thus universally assumed in the ancient Near East that lordship over Zaphon established the supremacy among the gods who could fly about. This, no doubt, was what Ba'al had expected. But seven years after the Fastness of Zaphon was completed, Ba'al faced a challenge by Mot, Lord of the southern lands and the Lower World. As it turned out, the dispute was no longer about the mastery of Zaphon; rather, it had to do with "who over the whole Earth dominion shall have."

Word somehow reached Mot, that Ba'al was engaged in suspicious activities. Unlawfully and clandestinely, he was "putting one lip to Earth and one lip to Heaven," and was attempting to "stretch his speech to the planets." Mot at first demanded the right to inspect the goings-on *within* the Crest of Zaphon. Instead, Ba'al sent emissaries with messages of peace. Who needs war? he asked; let us "pour peace and amity into the center of Earth." As Mot became more persistent, Ba'al concluded that the only way to prevent Mot from coming to Zaphon was for Ba'al to go to Mot's abode. So he journeyed to Mot's "pit" "in the depths of Earth," professing obedience.

Yet what he really had in mind was something more sinister—the overthrow of Mot. For that, he needed the help of the ever-faithful Anat. And so it was that while Ba'al had gone to Mot, his emissaries reached Anat at her own abode. The two emissaries were instructed to repeat to Anat, word for word, an enigmatic message:

> I have a word of secret to tell thee,
>   a message to whisper unto thee;
> It is a contraption that launches words,
>   a Stone that whispers.
> Men its messages will not know;
> Earth's multitudes will not comprehend.

"Stones" in the ancient languages, we must bear in mind, encompassed all quarried or mined substances and thus included all minerals and metals. Anat, therefore, readily understood what Ba'al was telling her: He was setting up upon the Crest of Zaphon some sophisticated contraption that could send or intercept secret messages!

The "Stone of Splendor" was further described in the secret message:

> Heaven with Earth it makes converse,
>   and the seas with the planets.
> It is a Stone of Splendor;
> To Heaven it is yet unknown.
> Let's you and I raise it
>   within my cavern, on lofty Zaphon.

So that was the secret: Ba'al, without the knowledge of "Heaven"—the government of the home planet—was setting up a clandestine communications center, from which he could converse with all the parts of Earth, as well as with the Spacecraft above Earth. It was a first step to "over the whole Earth dominion to have." In that, he ran into direct conflict with Mot; for it was in Mot's territories that the official "Eye of Earth" was located.

Having received and understood the message, Anat readily agreed to go to Ba'al's aid. The worried emissaries were promised that she would be there on time: "You are slow, I am swift," she assured them:

> The god's distant place I will penetrate,
>   the distant Hollow of the sons of the gods.
> Two openings it (has) under the Eye of Earth,
>   and three wide tunnels.

Arriving at Mot's capital, she could not find Ba'al. Demanding to know his whereabouts, she threatened Mot with violence. Finally, she learnt the truth: the two gods had engaged in hand combat, and "Ba'al was fallen." Enraged, she "with a sword cleaved Mot." Then with the aid of Shepesh, mistress of the Rephaim (the "Healers"), the lifeless body of Ba'al was flown back to the peak of Zaphon, and placed in a cavern.

Quickly, the two goddesses summoned the Craftsman of the Gods, also referred to as *El Kessem*—"The God of Magic." As Thoth had revived the snake-bitten Horus, so was Ba'al miraculously resurrected. But whether he was resurrected physically on Earth, or in a Celestial Afterlife (as Osiris), one cannot be certain.

When did the gods act out these events upon the Crest of Zaphon, no one can say for sure. But we do know that Mankind was cognizant of the existence and unique attributes of the "Landing Place" almost from the beginning of recorded history.

We have, to begin with, the journey of Gilgamesh to the Cedar Mountain, which his epic also calls "Abode of the gods, the Crossroads of Ishtar." There, "pressing into the forest," he came upon a tunnel which led to "the enclosure from which words of command are issued." Proceeding deeper into the mountain, "the secret abode of the Anunnaki he opened up." It is as though Gilgamesh had penetrated the very installations which Ba'al had secretly constructed! Mystery-filled verses in the epic now assume a thrilling meaningfulness:

> Secret things he has seen;
> What is hidden from Man, he knows. . . .

This, we know, took place in the third millenium B.C.—circa 2900 B.C.

The next link between the affairs of gods and men is the tale of the aging and heirless Danel, who had resided somewhere near Kadesh. There is no time frame given for the tale's occurrences, but the similarities with the biblical tale of the heirless Abraham—including the sudden appearance of "men" who turn out to be the Lord and his Emissaries, and the locale not far from Kadesh—suggest the possibility that we are reading two versions of the same ancestral memory. If so, we have another datemark: the beginning of the second millenium B.C.

Zaphon was still there, a Fastness of the Gods, in the first millenium B.C. The prophet Isaiah (eighth century B.C.) castigated the Assyrian invader of Judea, Sennacherib, for having insulted the Lord by ascending with his many chariots "to the mountain's height, unto the Crest of Zaphon." Stressing the antiquity of the place, the prophet transmitted to Sennacherib the Lord's admonition:

> Hast thou not heard it?
> Long ago have I made it,
> In days of old have I created it.

The same prophet likewise castigated the king of Babylon for having attempted to deify himself by scaling the Crest of Zaphon:

> O, how fallen from heaven art thou,
>   a Morning Star, son of Dawn!
> Felled to the ground
>   is he who the nations enfeebled.
> Thou didst say in thine heart
>   "I will ascend unto the heavens,
>   above the planets of El I shall raise my throne;
> On the Mount of Assembly I shall sit,
>   on the Crest of Zaphon.
> Upon the Raised Platform I shall go up,
>   a Lofty One I shall be!"
> But nay, to the Nether World you shall go,
>   down to the depths of a pit.

We have here not only confirmation of the existence of the place and its antiquity, but also a description of it: it included a "raised platform," from which one could ascend heaven-ward and become a "Lofty One"—a god.

The ascent heavenward, we know from other biblical writings, was by means of "stones" (mechanical contraptions) that could travel. In the sixth century B.C., the Prophet Ezekiel castigated the king of Tyre, whose heart grew haughty after he had been permitted to reach the Crest of Zaphon, and was taken within the "moving stones"—an experience after which he claimed "a god am I."

An ancient coin found at Byblos (the biblical Gebal), one of the Canaanite/Phoenician cities on the Mediterranean coast, may well have illustrated the structures erected upon Zaphon by Kothar-Hasis (Fig. 89). It depicts a "great house," adjoined by a raised area, which is surrounded by a high and massive wall. There, upon a podium supported by cross-beams built to withstand a great weight, there is mounted a conical object—an object familiar from so many other Near Eastern depictions: the Celestial Chamber of the gods—the "moving stone."

This, then, is the evidence bequeathed to us from antiquity. Millenium after millenium, the peoples of the ancient Near East were aware that within the Cedar Mountain there was

Fig. 89

a large platform for "moving stones," adjoined by a "great house" within which "a stone that whispers" was secreted.

And, if we have been right in our interpretations of ancient texts and drawings—could it be that this grand and known place had vanished?

IX

The Landing Place

The greatest Roman temple ruins lie not in Rome, but in the mountains of Lebanon. They encompass a grand temple to Jupiter—the grandest built anywhere in antiquity to honor any one god. Many Roman rulers, over a period of some four centuries, toiled to glorify this remote and alien place and erect its monumental structures. Emperors and generals came to it in search of oracles, to find out their fate. Roman legionnaires sought to be billeted near it; the devout and the curious went to see it with their own eyes: it was one of the wonders of the ancient world.

Daring European travelers, risking life and limb, reported the existence of the ruins since the visit there by Martin Baumgarten in January 1508. In 1751, the traveler Robert Wood, accompanied by the artist James Dawkins, restored some of the place's ancient fame when they described it in words and sketches. "When we compare the ruins . . . with those of many cities we visited in Italy, Greece, Egypt and other parts of Asia, we cannot help thinking them the remains of *the boldest plan* we ever saw attempted in architecture"— bolder in certain aspects than even the great pyramids of Egypt. The view upon which Wood and his companion had come was a panorama in which the mountaintop, the temples and the skies blended into one (Fig. 90).

The site is in the mountains of Lebanon, where they part to form a fertile, flat valley between the "Lebanon" range to the west and the "Anti-Lebanon" range to the east; where two rivers known from antiquity, the Litani and the Orontes,

Fig. 90

begin to flow into the Mediterranean Sea. The ruins were
of imposing Roman temples that were erected upon a vast
horizontal platform, artificially created at about 4,000 feet
above sea level. The sacred precinct was surrounded by a
wall, which served both as a retaining wall to hold the earthworks forming the flat top, as well as a fence to protect and
screen off the area. The enclosed squarish area, with some
sides almost 2,500 feet long, measured over five million
square feet.

Situated so as to command the flanking mountains and the
approaches to the valley from north and south, the sacred
area had its northwestern corner deliberately cut off—as
seen in this contemporary bird's eye view (Fig. 91a).

The right-angled cutout created an oblong area, which extended the platform's unimpeded northern view westward. It
was at that specially conceived corner that the vastest-ever
temple to Jupiter stood high, with some of the tallest (65
feet) and largest (7.5 feet in diameter) columns known in
antiquity. These columns supported an elaborately decorated superstructure ("architrave") 16 feet in height, atop
which there was a slanting roof, further raising the temple's
pinnacle.

The temple proper was only the westernmost (and oldest)

part of a four-part shrine to Jupiter, which the Romans are believed to have started to build soon after they occupied the place in 63 B.C.

Arranged along a slightly slanted east-west axis (Fig. 91b) were, first, a monumental Gateway ("A"); it comprised a grand staircase and a raised portico supported by twelve columns, in which there were twelve niches to hold the twelve Olympian gods. The worshippers then entered a forecourt ("B") of an hexagonal design, unique in Roman architecture; and through it continued to a vast altar court ("C"), which was dominated by an altar of monumental proportions: it rose some 60 feet from a base of about 70 by 70 feet. At the western end of the court stood the god's house proper ("D"). Measuring a colossal 300 by 175 feet, it stood upon a podium which was itself raised some 16 feet above the level of the court—a total of 42 feet above the level of the base platform. It was from that extra height that the tall columns, the architrave and the roof made together a real ancient skyscraper.

From its monumental gateway staircase to its final western wall, the shrine extended for more than 1,000 feet in length. It completely dwarfed a very large temple to its south ("E"), which was dedicated to a male deity, some think Bacchus but probably Mercury; and a small round temple ("F") to the southeast, where Venus was venerated. A German archaeological team that explored the site and studied its history on orders of Kaiser Wilhelm II, soon after he had visited the place in 1897, was able to reconstruct the layout of the sacred precinct and prepared an artist's rendering of what the ancient complex of temples, stairways, porticoes, gateways, columns, courtyards and altars probably looked like in Roman times (Fig. 92).

A comparison with the renowned *Acropolis* of Athens will give one a good idea of the scale of this Lebanese platform and its temples. The Athens complex (Fig. 93) is situated upon a stepped ship-like terrace less than 1,000 feet at its longest and about 400 feet at its widest. The stunning Par-

Fig. 91a

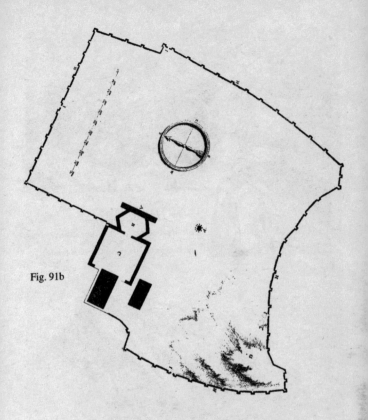

Fig. 91b

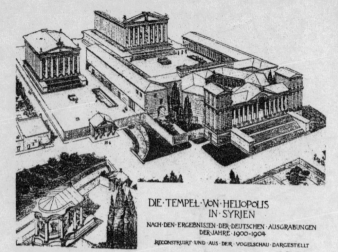

DIE · TEMPEL · VON · HELIOPOLIS
IN · SYRIEN

NACH · DEN · ERGEBNISSEN · DER · DEUTSCHEN · AUSGRABUNGEN
DER · JAHRE · 1900–1904

RECONSTRUIRT · UND · AUS · DER · VOGELSCHAU · DARGESTELLT

VON · BRUNO SCHULZ

Fig. 92

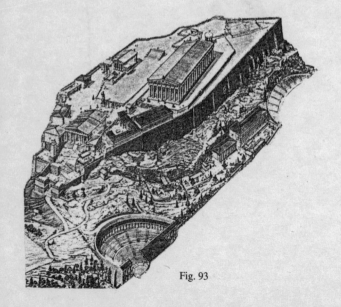

Fig. 93

thenon (temple of Athena) which still dominates the once sacred area and the whole plain of Athens is about 230 by 100 feet—even smaller than the temple of Mercury/Bacchus at the Lebanese Site.

Having visited the ruins, the archaeologist and architect Sir Mortimer Wheeler wrote two decades ago: "The temples . . . owe nothing of their quality to such new-fangled aids as concrete. They stand passively upon the *largest known stones in the world,* and some of their columns are *the tallest from antiquity.* . . . Here we have the last great monument . . . of the Hellenic world."

Hellenic world indeed, for there is no reason that any historian or archaeologist could find for this gigantic effort by the Romans, in an out-of-the-way place in an unimportant province, except for the fact that the place was hallowed by the Greeks who had preceded them. The gods to whom the three temples were dedicated—Jupiter, Venus and Mercury (or Bacchus)—were the Greek gods Zeus, his sister Aphrodite and his son Hermes (or Dionysus).

The Romans considered the site and its great temple as the ultimate attestations of the almightiness and supremacy of Jupiter. Calling him *Iove* (echo of the Hebrew *Yehovah*?), they inscribed upon the temple and its main statue the divine initials I.O.M.H.—the legend standing for *Iove Optimus Maximus Heliopolitanus:* the Optimal and Maximal Jupiter the Heliopolitan.

The latter title of Jupiter stemmed from the fact that though the great temple was dedicated to the Supreme God, the place itself was considered to have been a resting place of *Helios,* the Sun god who could traverse the skies in his swift chariot. The belief was transmitted to the Romans by the Greeks, from whom they also adopted the name of the place: *Heliopolis.* How the Greeks had come to so name the place, no one knows for sure; some suggest that it was so named by Alexander the Great.

Yet Greek veneration of the place must have been older and deeper rooted, for it made the Romans glorify the place with the greatest of monuments, and seek there the oracle's word concerning their fate. How else to explain the fact that, "in terms of sheer acreage, weight of stone, dimensions of the individual blocks, and the amount of carving, this precinct can scarcely have had a rival in the Graeco-Roman world" (John M. Cook, *The Greeks in Ionia and the East*).

In fact, the place and its association with certain gods go back to even earlier times. Archaeologists believe that there may have been as many as six temples built on the site before Roman times; and it is certain that whatever shrines the Greeks may have erected there, they—as the Romans after them—were only raising the structures atop earlier foundations, religiously and literally. Zeus (Jupiter to the Romans), it will be recalled, arrived in Crete from Phoenicia (today's Lebanon), swimming across the Mediterranean Sea after he had abducted the beautiful daughter of the king of Tyre. Aphrodite too came to Greece from western Asia. And the wandering Dionysus, to whom the second temple (or perhaps another) was dedicated, brought the vine and winemaking to Greece from the same lands of western Asia.

Aware of the worship's earlier roots, the Roman historian Macrobius enlightened his countrymen in the following words (*Saturnalia* I, Chapter 23):

> The Assyrians too worship the sun under the name of Jupiter, Zeus Helioupolites as they call him, with important rites in the city of Heliopolis. . . .
>
> That this divinity is at once Jupiter and the Sun is manifest both from the nature of its ritual and from its outward appearance. . . .
>
> To prevent any argument from ranging through a whole list of divinities, I will explain what the Assyrians believe concerning the power of the sun (god). They have given the name *Adad* to the god whom they venerate as highest and greatest. . . .

The hold the place had over the beliefs and imagination of people throughout the millenia also manifested itself in the history of the place following its Roman veneration. When Macrobius wrote the above, circa A.D. 400, Rome was already Christian and the site was already a target of zealous destruction. No sooner did Constantine the Great (A.D. 306–337) convert to Christianity, than he stopped all additional work there and instead began the conversion of the place into a Christian shrine. In the year 440, according to one chronicler, "Theodosius destroyed the temples of the Greeks; he transformed into a Christian church the temple of Heliopolis, that of *Ba'al Helios,* the great Sun-Ba'al of the celebrated Trilithon." Justinian (525–565) apparently carried off some of the pillars of red granite to Constantinople, the Byzantine capital, to build there the church of Hagia Sophia. These efforts to Christianize the place encountered repeated armed opposition by the local populace.

When the Muslims gained the area in the year 637, they converted the Roman temples and Christian churches atop the huge platform into a Muhammedan enclave. Where Zeus and Jupiter had been worshiped, a mosque was built to worship *Allah.*

Modern scholars have tried to shed more light on the age-long worship at this place by studying the archaeological evidence from neighboring sites. A principal one of these is Palmyra (the biblical Tadmor), an ancient caravan center on the way from Damascus to Mesopotamia. As a result, such scholars as Henry Seyrig *(La Triade Héliopolitaine)* and René Dussaud *(Temples et Cultes Héliopolitaine)* have concluded that a basic triad had been worshipped throughout the ages. It was headed by the God of the Thunderbolt and included the Warrior Maiden and the Celestial Charioteer. They and other scholars helped establish the now generally accepted conclusion, that the Roman-Greek triad stemmed from the earlier Semitic beliefs, which in turn were based upon the Sumerian pantheon. The earliest Triad was headed, it appears, by *Adad,* who was allotted

by his father Enlil—the chief god of Sumer—"the moun-
tainlands of the north." The female member of the Triad
was *Ishtar*. After he visited the area, Alexander the Great
struck a coin honoring Ishtar/Astarte and Adad; the coin
bears his name in Phoenician-Hebrew script (Fig. 94). The
third member of the Triad was the Celestial Charioteer,
Shamash—commander of the prehistoric astronauts. The
Greeks honored him (as Helios) by erecting a colossal
statue atop the main temple (see Fig. 92), showing him
driving his chariot. To them, its swiftness was denoted
by the four horses that pulled it; the authors of the *Book
of Enoch* knew better: "The chariot of Shamash," it says,
"was driven by the wind."

Fig. 94

Examining the Roman and Greek traditions and beliefs,
we have arrived back at Sumer; we have circled back to Gilgamesh and his Search for Immortality in the Cedar Forest, at the "crossroads of Ishtar." Though in the territory of
Adad, he was told, the place was also within the jurisdiction
of Shamash. And so we have the original Triad: Adad, Ishtar,
Shamash.

Have we come upon the Landing Place?

That the Greeks were aware of the epic adventures of Gilgamesh, few scholars nowadays doubt. In their "investigation of the origins of human knowledge and its transmission

through Myth," entitled *Hamlet's Mill,* Giorgio de Santillana and Hertha von Deschend point out that "Alexander was a true replica of Gilgamesh." But even earlier, in the historic tales of Homer, the heroic Odysseus had already followed similar footsteps. Shipwrecked after traveling to the abode of Hades in the Lower World, his men reached a place where they "ate the cattle of the Sun god" and were therefore killed by Zeus. Left alive alone, Odysseus wandered about until he reached the "Ogygian island"—the secluded place from pre-Deluge times. There, the goddess Calypso, "who kept him in a cave and fed him, wanted him to marry her; in which case she intended making him immortal, so that he should never grow old." But Odysseus refused her advances—just as Gilgamesh had turned down Ishtar's offer of love.

Henry Seyrig, who as Director of Antiquities of Syria devoted a lifetime to the study of the vast platform and its meaning, found that the Greeks used to conduct there "rites of mystery, in which Afterlife was represented as human Immortality—an identification with the deity obtained by the ascent (heavenward) of the soul." The Greeks, he concluded, indeed associated this place with Man's efforts to attain Immortality.

Was then this place the very place in the Cedar Mountains to which Gilgamesh had first gone with Enkidu, the Crest of Zaphon of Ba'al?

To give a definite answer, let us look more closely at the physical features of the place. We will find that the Romans and Greeks have built their temples upon a paved platform which existed from much earlier times—a platform constructed of large, thick stone blocks so tightly put together that no one—to this very day—has been able to penetrate it and study the chambers, tunnels, caverns and other substructures that lie hidden beneath it.

That such subterranean structures undoubtedly exist is judged not only from the fact that other Greek temples had secret, subterranean cellars and grottoes beneath their apparent floors. Georg Ebers and Hermann Guthe (*Palästina in Bild und Wort;* the English version is titled *Picturesque Palestine*) reported a century ago that the local Arabs entered the ruins "at the southeast corner, through a long vaulted passage like a railway tunnel *under the great platform*" (Fig. 95). "Two of these great vaults run parallel with each other, from east to west, and are connected by a third running at right angles to them from north to south." As soon as they entered the tunnel, they were caught in total darkness, broken here and there by eerie green lights from puzzling "laced windows." Emerging from the 460-feet-long tunnel, they found themselves under the north wall

Fig. 95

of the Sun Temple, "which the Arabs call *Dar-as-saadi*—House of Supreme Blissfulness."

The German archaeologists also reported that the platform apparently rested upon gigantic vaults; but they concerned themselves with mapping and reconstructing the superstructure. A French archaeological mission, led by André Parrot in the 1920s, confirmed the existence of the subterranean maze, but was unable to penetrate its hidden parts. When the platform was pierced from above through its thick stones, evidence was found of structures beneath it.

The temples were erected upon a platform raised to thirty feet, depending on the terrain. It was paved with stones whose size, to judge by the slabs visible at the edges, ranged from a length of twelve feet to thirty feet, a frequent width of nine feet and a thickness of six feet. No one has yet attempted to calculate the quantity of stone hewn, cut, shaped, hauled and imbedded layer upon layer upon this site; it could possibly dwarf the Great Pyramid of Egypt.

Whoever laid this platform originally, paid particular attention to the rectangular northwestern corner, the location of the temple of Jupiter/Zeus. There, the temple's more than 50,000 square feet rested upon a raised podium which was certainly intended to support some extremely heavy weight. Constructed of layer upon layer of huge stones, the Podium rose twenty-six feet above the level of the Court in front of it and forty-two feet above the ground on its exposed northern and western sides. On the southern side, where six of the temple's columns still stand, one can clearly see (Fig. 96a) the stone layers: interspersed between sizable yet relatively small stones, there are alternating rows of stone blocks measuring up to twenty-one feet in length. One can also see (bottom left) the lower layers of the Podium, protruding as a terrace below the raised temple. There, the stones are even more gigantic.

More massive by far were the stone blocks in the western side of the Podium. As shown in the schematic drawing of the northwestern corner prepared by the German archaeological team (Fig. 96b), the protruding base and the top layers of the Podium were constructed of "cyclopian" stone blocks some of which measure over thirty-one feet in length, about thirteen feet in height and are twelve feet thick. Each such slab represents thus about 5,000 cubic feet of stone and weighs more than 500 tons.

Large as these stones are—the largest ones in the Great

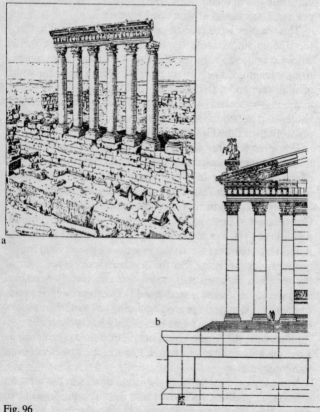

Fig. 96

Pyramid of Egypt weigh 200 tons—they were not the largest slabs of granite employed by the ancient master builder in creating the Podium.

The central layer—situated some twenty feet above the base of the Podium—was incredibly made up of even larger stones. Modern surveyors have spoken of them as "giant," "colossal," "huge." Ancient historians named them the *Trilithon*—the Marvel of the Three Stones. For there, exposed to view in the western side of the Podium, lie side by side three stone blocks, the likes of which cannot be seen anywhere else in the world. Precisely shaped and perfectly fitting, each of the three stones (Fig. 97) measures over sixty feet in length, with sides of fourteen and twelve feet. Each slab thus represents more than 10,000 cubic feet of granite and weighs well over 1,000 tons!

The stones for the Platform and Podium were quarried locally; Wood and Dawkins include one of these quarries (Fig. 90) in their panoramic sketch, showing some of the large stone blocks strewn around in the ancient quarry. But the gigantic blocks were hewn, cut and shaped at another quarry, situated in the valley some three-quarters of a mile southwest of the sacred precinct. It is there that one comes upon a sight even more incredible than that of the Trilithon.

Partly buried in the ground, there lies yet another one of the colossal granite slabs—left *in situ* by whoever the grand quarrier was. Fully shaped and perfectly cut, with only a thin line at its base still connecting it to the rocky ground, it is an unbelievable sixty-nine feet long and has a girth of sixteen by fourteen feet. A person climbing upon it (Fig. 98) looks like a fly upon a block of ice. . . . It weighs, by conservative estimates, more than 1,200 tons.

Most scholars believe that it was intended to be hauled, as its three sisters were, up to the sacred precinct and perhaps be used to extend the terrace-part of the Podium on the northern side. Ebers and Guthe record a theory that in the row beneath the Trilithon, there are not two smaller slabs but a single stone akin to the one found at the quarry, measuring

Fig. 97

Fig. 98

more than sixty-seven feet in length, but either damaged or otherwise chiseled to give the appearance of two side-by-side stones.

Wherever the leftover colossal stone was intended to be placed, it serves as a mute witness to the immensity and uniqueness of the Platform and Podium nesting in the mountains of Lebanon. The mind-boggling fact is that even nowadays there exists no crane, vehicle or mechanism that can lift such a weight of 1,000–1,200 tons—to say nothing of carrying such an immense object over valley and mountainside, and placing each slab in its precise position, many feet above the ground. There are no traces of any roadway, causeway, ramp or other earthworks that could even remotely suggest the hauling or dragging of these megaliths from the quarry to their uphill site.

Yet in remote days, someone, somehow had achieved the feat. . . .

But who? Local traditions hold that the place had existed from the days of Adam and his sons, who resided in the area of the Cedar Mountains after the expulsion of Adam and Eve from the Garden of Eden. Adam, these legends relate, inhabited the place which is now Damascus, and died not far from there. It was Cain his son who built a refuge upon the Cedar Crest after he had killed Abel.

The Maronite Patriarch of Lebanon related the following tradition: "The fastness on Mount Lebanon is the most ancient building in the world. Cain, the son of Adam, built it in the year 133 of Creation, during a fit of raving madness. He gave it the name of his son Enoch, and peopled it with giants who were punished for their iniquities by the Flood." After the Deluge, the place was rebuilt by the biblical Nimrod, in his efforts to scale the heavens. The Tower of Babel, according to these legends, was not in Babylon but upon the great platform in Lebanon.

A seventeenth century traveler named d'Arvieux wrote in his *Mémoires* (Part II, Chapter 26) that local Jewish inhabitants, as well as Muslim residents, held that an ancient manuscript found at the site revealed that "After the Flood, when Nimrod reigned over Lebanon, he sent giants to rebuild the Fortress of Baalbek, which is so named in honor of Ba'al, the god of the Moabites, worshippers of the Sun-god."

The association of the god Ba'al with the place in post-Diluvial days strikes a bell. Indeed, no sooner were the Greeks and Romans gone, than the local people abandoned the Hellenistic name Heliopolis and resumed calling the place by its Semitic name. It is the name by which it is still called to this day: *Baalbek.*

There are differing opinions as to the precise meaning of the name. Many believe that it means "The Valley of Ba'al." But from the spelling and from Talmudic references, we surmise that it has meant *"The Weeping of Ba'al."*

We can hear again the closing verses of the Ugaritic epic, describing the fall of Ba'al in his struggle with Mot, the discovery of his lifeless body, his entombment by Anat and Shepesh in a grotto upon the Crest of Zaphon:

> They came upon Baal, fallen on the ground;
> Puissant Baal is dead;
> The Prince, Lord of Earth, is perished. . . .
> Anat weeps her fill of weeping;
> In the valley she drinks tears like wine.
> Loudly she calls unto the Gods' Torch, Shepesh:
> "Lift Puissant Baal, I pray,
>   lift him onto me."
> Hearkening, the Gods' Torch Shepesh
> Picks up Puissant Baal,
> Sets him on Anat's shoulder.
> Up to the Fastness of Zaphon she brings him;
> Bewails him, entombs him;
> Lays him in the hollows of the earth.

All these local legends, which as all legends contain a kernel of age-old recollections of actual events, agree that the place is of extreme antiquity. They ascribe its building to "giants" and connect its construction with the events of the Deluge. They connect it with Ba'al, its function being that of a "Tower of Babel"—a place from which to "scale the heavens."

As we look at the vast Platform, its location and layout, and ponder the purpose of the immense Podium built to sustain massive weights, the depiction on the coin from Byblos (Fig. 89) keeps flashing before our eyes: a great temple, a walled sacred area, a podium of extra-strong construction—and upon it the rocket-like Flying Chamber.

The words and descriptions of the Hidden Place in the Epic of Gilgamesh also keep echoing in our ears. The insurmountable wall, the gate which stuns whoever touches it, the tunnel to the "enclosure from which words of command are issued," the "secret abode of the Anunnaki," the monstrous Guardian with his "radiant beam."

And there is no doubt left in our mind that in *Baalbek* we have found Ba'al's Crest of Zaphon, the target of the first journey of Gilgamesh.

The designation of Baalbek as "the Crossroads of Ishtar" implies that, as she roamed Earth's skies, she could come and go from that "Landing Place" to other landing places upon Earth. Likewise, the attempt by Ba'al to install upon the Crest of Zaphon "a contraption that launches words, a 'stone that whispers,'" implied the existence elsewhere of similar communication units: "Heaven with Earth it makes converse, and the seas with the planets."

Were there indeed such other places on Earth that could serve as Landing Places for the aircraft of the gods? Were there, besides upon the Crest of Zaphon, other "stones that whisper"?

The first obvious clue is the very name "Heliopolis," indicating the Greek belief that Baalbek was, somehow, a "City of the sun god" paralleling its namesake city in Egypt. The Old Testament too recognized the existence of a northern

Beth-Shemesh ("House of Shamash") and a southern Beth-Shemesh, *On,* the biblical name for the Egyptian Heliopolis. It was, the prophet Jeremiah said, the place of the "Houses of the gods of Egypt," the location of Egypt's obelisks.

The northern Beth-Shemesh was in Lebanon, not far from *Beth-Anath* ("House/Home of Anat"); the prophet Amos identified it as the location of the "palaces of Adad . . . the House of the one who saw El." During the reign of Solomon, his domains encompassed large parts of Syria and Lebanon, and the list of places where he had built great structures included *Baalat* ("The Place of Ba'al") and *Tamar* ("The Place of Palms"); most scholars identify these places as Baalbek and Palmyra (see map, Fig. 78).

Greek and Roman historians made many references to the links that connected the two Heliopolises. Explaining the Egyptian pantheon of twelve gods to his countrymen, the Greek historian Herodotus also wrote of an "Immortal whom the Egyptians venerated as 'Hercules.'" He traced the origins of the worship of this Immortal to Phoenicia, "hearing that there was a temple of Hercules at that place, very highly venerated." In the temple he saw two pillars. "One was of pure gold; the other was of emerald, shining with great brilliancy at night."

Such sacred "Sun Pillars"—"Stones of the gods"—were actually depicted on Phoenician coins following the area's conquest by Alexander (Fig. 99). Herodotus provides us with the additional information that of the two connected stones, one was made of the metal which is the best conductor of electricity (gold); and the other of a precious stone (emerald) as is now used for laser communications, giving off an eerie radiance as it emits a high-powered beam. Was it not like the contraption set up by Ba'al, which the Canaanite text described as "stones of splendor?"

The Roman historian Macrobius, writing explicitly about the connection between the Phoenician Heliopolis (Baalbek) and its Egyptian counterpart, also mentions a sacred stone; according to him, "an object" venerating the Sun god

Zeus Helioupolites was carried by Egyptian priests from the Egyptian Heliopolis to Heliopolis (Baalbek) in the north. "The object," he added, "is now worshipped with Assyrian rather than Egyptian rites."

Other Roman historians also stressed that the "sacred stones" worshiped by the "Assyrians" and the Egyptians were of a conical shape. Quintus Curtius recorded that such an object was located at the temple of Ammon at the oasis of Siwa. "The thing which is worshipped there as a god," Quintus Curtius wrote, "has not the shape that artificers have usually applied to the gods. Rather, its appearance is most like an *umbilicus,* and it is made of an emerald and gems cemented together."

The information regarding the conical object worshiped at Siwa was quoted by F. L. Griffith in connection with the announcement, in *The Journal of Egyptian Archaeology* (1916), of the discovery of a conical "omphalos" at the Nubian "pyramid city" of Napata. This "unique Meroitic monument" (Fig. 100) was found by George A. Reisner of Harvard University at the inner sanctum of the temple of Ammon there—the southernmost temple to this god of Egypt.

The term *omphalos* in Greek or *umbilicus* in Latin means a "navel"—a conical stone which, for reasons that scholars do not understand, was deemed in antiquity to have marked a "center of the Earth."

Fig. 99

Fig. 100

The temple of Ammon at the oasis of Siwa, it will be recalled, was the location of the oracle which Alexander rushed to consult on his arrival in Egypt. We have the testimony of both Callisthenes, Alexander's historian, and the Roman Quintus Curtius that an omphalos made of precious stones was the very "object" venerated at that oracle site. The Nubian temple of Ammon where Reisner discovered the omphalos stone was at Napata, an ancient capital of the domains of Nubian queens; and we recall the baffling visit of Alexander to Queen Candace, in his continuing quest for Immortality.

Was it mere coincidence that, in his search for the secrets of longevity, the Persian king Cambyses (as Herodotus has reported) sent his men to Nubia, to the temple where the "Table of the Sun" was enshrined? Early in the first millenium B.C. a Nubian queen—the Queen of Sheba—made a long journey to King Solomon in Jerusalem. The legends current at Baalbek relate that he embellished the site in Lebanon in her honor. Did she then undertake the long and hazardous voyage merely to enjoy the wisdom of Solomon, or was her real purpose to consult the oracle at Baalbek—the biblical "House of Shemesh?"

There seem to be more than just coincidences here; and the question that comes to mind is this: if at all these oracle centers an omphalos was enshrined—was the omphalos itself the very source of the oracles?

The construction (or reconstruction) upon the Crest of Zaphon of a launching silo and a landing platform for Ba'al was not the cause of his fatal battle with Mot. Rather, it was his clandestine attempt to set up a "Stone of Splendor." This device could communicate with the heavens as well as with other places on Earth. But, in addition, it was

> A stone that whispers;
> Men its messages will not know,
> Earth's multitudes will not comprehend.

As we ponder the apparent dual function of the Stone of Splendor, the secret message of Ba'al to Anat all of a sudden becomes clear: the same device which the gods used to communicate with each other was also the object from which there emanated the gods' oracular answers to the kings and heroes!

In a most thorough study on the subject, Wilhelm H. Roscher *(Omphalos)* showed that the Indo-European term for these oracle stones—*navel* in English, *nabel* in German, etc.—stem from the Sanskrit *nabh,* which meant "emanate forcefully." It is no coincidence that in the Semitic languages *naboh* meant to foretell and *nabih* meant "prophet." All these identical meanings undoubtedly harken back to the Sumerian, in which NA.BA(R) meant "bright-shiny stone that solves."

A veritable network of such oracle sites emerges as we study ancient writings. Herodotus—who accurately reported (Book II, 29) the existence of the Meroitic oracle of Jupiter-Ammon—added to the links we have so far discussed by stating that the "Phoenicians," who established the oracle at Siwa, also established the oldest oracle center in Greece, the one at *Dodona* —a mountain site in northwestern Greece (near the present Albanian border).

To that effect, he related a report he had heard when he visited Egypt, whereby "two sacred women were once carried off from Thebes (in Egypt) by the Phoenicians . . . one of them was sold into Libya (western Egypt) and the other into Greece. These women were the first founders of the oracles in the two countries." This version, Herodotus wrote, he heard from the Egyptian priests of Thebes. But at Dodona, the version was that "two black doves flew away from the Egyptian Thebes," one alighting at Dodona and the other at Siwa: whereupon an oracle of Jupiter was established at both places, the Greeks calling him Zeus at Dodona and the Egyptians Ammon at Siwa.

The Roman historian Silicus Italicus (first century A.D.),

relating that Hannibal consulted the oracle at Siwa regarding his wars against Rome, also credited the flight of the two doves from Thebes with the establishment of the oracles in the Libyan desert (Siwa) and in Greek Chaonia (Dodona). Several centuries later, the Greek poet Nonnos, in his master work *Dionysiaca*, described the oracle shrines at Siwa and Dodona as twin sites, and held that the two were in voice communication with each other:

> Behold the new-found answering voice
>   of the Libyan Zeus!
> The thirsty sands an oracular sent froth
>   to the dove at Chaonia [= Dodona].

As far as F. L. Griffith was concerned, the discovery of the omphalos in Nubia brought to mind another oracle center in Greece. The conical shape of the Nubian omphalos, he wrote, "was precisely that of the omphalos at the oracle at *Delphi*."

Delphi, the site of Greece's most famous oracle, was dedicated to Apollo ("He of Stone"); its ruins are still one of Greece's leading tourist attractions. There too, as at Baalbek, the sacred precinct consisted of a platform shaped upon a mountainside, also facing a valley that opens up as a funnel toward the Mediterranean Sea and the lands on its other shores.

Many records establish that an omphalos stone was Delphi's holiest object. It was set into a special base in the inner sanctum of the temple of Apollo, some say next to a golden statue of the god and some say it was enshrined all by itself. In a subterranean chamber, hidden from view by the oracle seekers, the oracle priestess, in trance-like oblivion, answered the questions of kings and heroes by uttering enigmatic answers—answers given by the god but emanating from the omphalos.

The original sacred omphalos had mysteriously disappeared, perhaps during the several sacred wars or foreign invasions which affected the place. But a stone replica thereof, erected perhaps in Roman times outside the temple, was discovered in archaeological excavations and is now on display in the Delphi Museum (Fig. 101).

Along the Sacred Way leading up to the temple, someone, at some unknown time, also set up a simple stone omphalos in an effort to mark the place where oracles were first given at Delphi, before the temple was built. The coins of Delphi depicted Apollo seated on this omphalos (Fig. 102); and after Phoenicia fell to the Greeks, they likewise depicted Apollo seated upon the "Assyrian" omphalos. But just as frequently, the oracle stones were depicted as twin cones connected to each other via a common base, as in Fig. 99.

Fig. 101

How was Delphi chosen as a sacred oracle place, and how did the omphalos stone come to be there? The traditions say that when Zeus wanted to find the center of the Earth, he released eagles from two opposite ends of the world. Flying toward each other, they met at Delphi; whereupon the place was marked by erecting there a navel stone, an omphalos. According to the Greek historian Strabo, images of two such eagles were perched on top of the omphalos at Delphi.

Depictions of the omphalos have been found in Greek art, showing the two birds atop or at the sides (Fig. 102) of the

Fig. 102

conical object. Some scholars see in the birds not eagles, but
carrier pigeons, which—being able to find their way back to
a certain place—might have symbolized the measuring of
distances from one Center of Earth to another.

According to Greek legends, Zeus found refuge at Delphi during his aerial battles with Typhon, resting on the
platform-like area upon which the temple to Apollo was
eventually built. The shrine to Ammon at Siwa contained not
only subterranean corridors, mysterious tunnels and secret
passages inside the temple's thick walls, but also a restricted
area of some 180 by 170 feet, surrounded by a massive wall.
In its midst, there arose a solid stone platform. We find the
same structural components, including a raised platform, in

all the sites associated with the "stones that whisper." Is one to conclude, then, that as the far larger Baalbek was, they too were both a Landing Place and a Communications Center?

Not surprisingly, we find the twin Sacred Stones, accompanied by the two eagles, also depicted in Egyptian sacred writings (Fig. 103); and many centuries before the Greeks even began to enshrine their oracle centers, an Egyptian Pharaoh depicted an omphalos with the two perched birds in his pyramids. He was Seti I, who lived in the fourteenth century B.C.; and it was in his depiction of the domain of Seker, the Hidden God, that we have seen the oldest omphalos to date—in Fig. 19. It was the communications means whereby messages—"words"—"were spoken to Seker every day."

In Baalbek, we have found the target of the first journey of Gilgamesh. Having followed the threads connecting the "whispering" Stones of Splendor, we arrived at the *Duat*.

It was the place where the Pharaohs sought the Stairway to Heaven for an Afterlife. It was, we suggest, the place whereto Gilgamesh, in search of Life, set his course on his second journey.

Fig. 103

X

Tilmun: Land of the Rocketships

The epic search of Gilgamesh for Immortality has undoubtedly been the fountainhead of the many tales and legends, in subsequent millenia, of kings and heroes who have likewise gone to find everlasting youth. Somewhere on Earth, Mankind's mythified memories held, there was a place where Man could join the gods, and be spared the indignity of death.

Nearly 5,000 years ago, Gilgamesh of Uruk had pleaded with Utu (Shamash):

> In my city, man dies; oppressed is my heart.
> Man perishes; heavy is my heart . . .
> Man, the tallest, cannot stretch to Heaven . . .
> O Utu,
> The Land I wish to enter; be thou my ally . . .
> In the place where the *Shems* have been raised up,
> Let me set up my *Shem*!

The *Shem,* we have shown, though commonly translated "Name" (that by which one is remembered), was in fact a rocketship: Enoch vanished upon his "Name" as he was taken heavenward. Half a millenium after Gilgamesh, in Egypt, King Teti made an almost identical plea:

> Men fall,
> They have no *Name*.

(O god),
Seize thou King Teti by his arms,
Take thou King Teti to the sky,
That he die not on Earth among men.

The goal of Gilgamesh was Tilmun, the land where the rocketships were raised up. To ask where he went to reach Tilmun, is to ask where Alexander went, deeming himself a Pharaoh and a god's son. It is to ask: Where on Earth was the *Duat?*

Because all these destinations, we must conclude, were one and the same.

And the land where they hoped to find the Stairway to Heaven, we shall conclusively show, was the peninsula of Sinai.

Accepting the possibility that the details given in the Book of the Dead may indeed refer to actual Egyptian geography, some scholars have suggested that the Pharaoh's simulated journey was along the Nile, from the shrines in Upper Egypt to those in Lower Egypt. The ancient texts, however, clearly speak of a journey beyond the boundaries of Egypt. The Pharaoh's direction is eastward, not northward; and as he crosses the Lake of Reeds and the desert beyond it, he leaves behind not only Egypt but also Africa: much is made of the perils—real and "political"—of coming from the domains of Horus into the "Lands of Seth," to Asia.

When the Pyramid Texts were inscribed by the Pharaohs of the Old Kingdom, their capital in Egypt was Memphis. The religious center of old was Heliopolis, a short distance to the northeast of Memphis. From these centers, a course eastward in fact led to a chain of lakes full of reeds and rushes. Beyond lay the desert, and the mountain pass, and the Sinai peninsula—the area whose skies had served as the final battlefield between Horus and Seth, between Zeus and Typhon.

The suggestion that the Pharaoh's journey to the Afterlife had indeed taken him to the Sinai peninsula is supported by the fact that Alexander had emulated not only the Pharaohs; there was also a deliberate effort to emulate the Israelite Exodus from Egypt under the leadership of Moses.

As in the biblical tale, the starting point was Egypt. Next came the "Red Sea"—the watery barrier whose waters parted so that the Israelites could cross upon the dry bed. In the histories of Alexander, the watery barrier was also encountered, and it was persistently called the Red Sea. As in the tale of Exodus, Alexander too attempted to lead his army across the waters on foot: in one version by building a causeway; in another "Alexander lay it bare by his prayers." Whether he succeeded or not (depending on the version), enemy soldiers were drowned by the onrushing waters—just as the Egyptians pursuing the Israelites had been drowned. The journeying Israelites encountered and battled an enemy named the Amalekites: in a Christian version of the Alexander histories, the enemy destroyed "by means of collecting the waters of the Red Sea and pouring the waters over them" were called "Amalekites."

Once across the waters—the literal translation of the biblical term *Yam Suff* is "Sea/Lake of Reeds"—there began a journey in a desert, toward a sacred mountain. Significantly, the landmark mountain which Alexander reached was named *Mushas*—the Mountain of Moses, whose Hebrew name was *Moshe.* It was there that Moses encountered an angel who spoke to him out of a fire (the burning bush); a similar incident is described in the tales of Alexander.

The double and triple parallels multiply, as we recall the tale in the Koran of Moses and the fish. The location of the Waters of Life in the Koran tale about Moses was "the junction of the two streams." It was where the stream of Osiris divided into two tributaries, that the Pharaoh's reached the entrance to the subterranean realm. In the tales of Alexander, it was at the junction of two subterranean streams that the crucial point was reached, where the "Stone of Adam" emitted light, where Alexander was advised by divine beings to turn back.

And there was the tradition, also recorded in the Muslim Koran, equating Alexander with Moses by calling him "He of the Two Horns"—recalling the biblical statement that, after Moses had visited with the Lord upon Mount Sinai, his face radiated and emitted "horns" (literally: rays) of light.

The arena for the biblical Exodus was the peninsula of Sinai. The conclusion from all the similarities and footstep-following can only be that it was toward the Sinai peninsula that Alexander, Moses and the Pharaohs set their course as they went east from Egypt. This, we will show, was also the destination of Gilgamesh.

To reach *Tilmun* on his second and decisive journey, Gilgamesh set sail in a "Ship of *Magan*," a Ship of Egypt. Starting from Mesopotamia, his only course was to sail down the Persian Gulf. Then, rounding the Arabian peninsula, he would have entered the Red Sea (which the Egyptians called the Sea of Ur). As the name of his ship indicates, he would have sailed up the Red Sea toward Egypt. But his destination was not Egypt; it was Tilmun. Was he then intending to land on the western shores of the Red Sea—in Nubia? On the eastern shore—in Arabia? Or straight ahead, on the peninsula of Sinai? (See map, Fig. 2.)

Fortunately for our investigation, Gilgamesh had met with a misfortune. His ship was sunk by a guarding god soon after he began his voyage. He was not too far gone from Sumer, for Enkidu (whose presence on the ship caused its sinking) pleaded that they make their way back, on foot, to Uruk. Resolved to reach Tilmun, Gilgamesh trekked instead over-land to his chosen destination. Were his goal on the shores of the Red Sea, he would have cut across the Arabian penin-sula. But instead he set his course to the northwest. We know that for a fact, because—having crossed a desert and passed desolate mountains—his first glimpse of civilization was a "low-lying sea." There was a city nearby, and an inn on its outskirts. The "ale-woman" warned him that the sea he saw and wished to cross was the "Sea of the Waters of Death."

Just as the Cedars of Lebanon had served as a unique

landmark for determining the first destination of Gilgamesh, so does the "Sea of the Waters of Death" serve as a unique clue to the whereabouts of Gilgamesh on his second journey. Throughout the Near East, in all the lands of the ancient world, there is only one such body of water. It is so called to this very day: the *Dead Sea*. It is, indeed, a "low-lying sea," being the lowest body of water on the face of Earth (1,300 feet below sea level). Its waters are so saturated with salts and minerals that it is totally devoid of all marine and plant life.

The city that overlooked the Sea of the Waters of Death was surrounded by a wall. Its temple was dedicated to Sin, the Moon-god. Outside the city there was an inn. The hostess took Gilgamesh in, extending to him hospitality, giving him information.

The uncanny similarities to a known biblical tale cannot be missed. When the Israelites' forty years of wandering in the Wilderness had come to an end, it was time to enter Canaan. Coming from the Sinai peninsula, they circled the Dead Sea on its eastern side until they reached the place where the Jordan River flows into the Dead Sea. When Moses stood upon a hill overlooking the plain, he could see—as Gilgamesh had seen—the shimmering waters of the "low-lying sea." In the plain, on the other side of the Jordan, stood a city: *Jericho*. It blocked the Israelites' advance into Canaan, and they sent two spies to explore its defenses. A woman whose inn was at the city's walls extended to them hospitality, gave them information and guidance.

The Hebrew name for Jericho is *Yeriho*. It literally means "Moon City"—the city dedicated to the Moon god, Sin. . . .

It was, we suggest, the very city reached by Gilgamesh fifteen centuries before the Exodus.

Was Jericho already in existence circa 2900 B.C., when Gilgamesh was engaged in his searches? Archaeologists are agreed that Jericho has been inhabited since before 7000 B.C., and served as a flourishing urban center since about 3500 B.C.; it was certainly there when Gilgamesh arrived.

Refreshed and back to strength, Gilgamesh planned his continued journey. Finding himself at the northern end of the Dead Sea, he inquired of the ale-woman whether he could sail across its waters, rather than circle it overland. Were he to take the overland route, he would have taken the route which the Israelites eventually took—but in reverse; for Gilgamesh wished to go where the Israelites eventually came from. When the boatman Urshanabi finally ferried him over, he stepped ashore, we believe, at the southern end of the Dead Sea—as close to the Sinai peninsula as the boatman could have taken him.

From there he was to follow "a regular way"—a route in common use by caravans—"toward the Great Sea, which is far away." Again, the geography is recognizable from biblical terminology, for the Great Sea was the biblical name for the Mediterranean Sea. Journeying in the Negev, the dry southern region of Canaan, Gilgamesh was to go westward for a certain distance, looking for "two stone markers." There, Urshanabi told him, he was to make a turn and reach the town named Itla; it was located some distance from the Great Sea. Beyond Itla, in the Fourth Region of the gods, lay the restricted area.

Was Itla a "City of the Gods" or a City of Men?

The events there, described in a fragmented Hittite version of the Gilgamesh Epic, indicate that it was a place for both. It was a "sanctified city," with various gods coming and going through it or within easy reach of it. But men too could go there: the way to it was indicated by road markers. Gilgamesh not only rested there, and changed into fresh clothing: he also obtained there the sheep which he daily offered as sacrifices to the gods.

Such a city is known to us from the Old Testament. It was located where the south of Canaan merges into the Sinai peninsula, a gateway into the peninsula's Central Plain. Its sanctity was denoted by its name: *Kadesh* ("The Sanctified"); it was distinguished from a northern namesake (situated, significantly, on the approaches to Baalbek) by being called

Kadesh-Barnea (which, stemming from the Sumerian, could have meant "Kadesh of the Shiny Stone Pillars"). In the Age of the Patriarchs, it was included in the domain of Abraham, who "journeyed to the Negev, and dwelt between Kadesh and Shur."

The city, by name and by function, is also known to us from the Canaanite tales of gods, men and the craving for Immortality. Danel, we recall, asked the god El for a rightful heir, so that his son could erect for him a commemorative stela at Kadesh. In another Ugaritic text we are told that a son of El named *Shibani* ("The Seventh"),—the biblical town of Beer-Sheba ("The Well of the Seventh") might have been named after him—was told to "raise a commemorating (Pillar) in the desert of Kadesh."

Indeed, both Charles Virolleaud and René Dussaud, who in the periodical *Syria* pioneered the translation and understanding of the Ugaritic texts, concluded that the locale of the many epic tales "was the region between the Red Sea and the Mediterranean," the Sinai peninsula. The god Ba'al, who loved to fish in Lake Sumkhi, went for his hunting to the "desert of Alosh," an area associated (as in Fig. 104) with the date palm. As both Virolleaud and Dussaud have pointed out, this is a geographical clue connecting the Ugaritic locale with the biblical record of the Exodus: the Israelites, according to Numbers 33, journeyed from Marah (the place of bitter waters) and Elim (the oasis of date palms) to *Alosh*.

Fig. 104

More details, placing El and the younger gods in the same arena as that of the Exodus, are found in a text entitled by the scholars "The Birth of the Gracious and Beautiful Gods." Its very opening verses locate the action in the "Desert of *Suffim*"—unmistakably a desert bordering on the *Yam Suff* ("Sea of Reeds") of the Exodus:

> I call the gracious and beautiful gods,
>> sons of the Prince. I will place them
>> in the City of Ascending and Going,
>> in the desert of Suffim.

The Canaanite texts provide us with yet another clue. By and large they refer to the pantheon's head as "El"—the supreme, the loftiest—a generic title rather than a personal name. But in the above quoted text El identifies himself as *Yerah* and his spouse as *Nikhal*. "Yerah" is the Semitic for "Moon"—the god better known as *Sin;* and "Nikhal" is a Semitic rendition of NIN.GAL, the Sumerian name for the spouse of the Moon-god.

Scholars have advanced many theories regarding the origin of the peninsula's name *Sinai*. For once, the obvious reason—that, as the name stated, it "Belonged to *Sin*"—has been among the preferred solutions.

We can see (in Fig. 72) that the Moon's crescent was the emblem of the deity in whose land the Winged Gateway was located. We find that the main crossroads in the central Sinai, the well-watered place *Nakhl,* still bears the name of Sin's spouse.

And we can confidently conclude that the "Land Tilmun" was the Sinai peninsula.

An examination of the geography, topography, geology, climate, flora and history of the Sinai peninsula will affirm our identification, and clarify the Sinai's role in the affairs of gods and men.

The Mesopotamian texts described Tilmun as situated at the "mouth" of two bodies of water. The Sinai peninsula, shaped as an inverted triangle indeed begins where the Red Sea separates into two arms—the Gulf of Suez on the west, and the Gulf of Elat (Gulf of Aqaba) on the east. Indeed, when Egyptian depictions of the Land of Seth, where the Duat was, are turned around, they show schematically a peninsula with the Sinai's features (Fig. 105).

The texts spoke of "mountainous Tilmun." The Sinai peninsula is indeed made up of a high mountainous southern part, a mountainous central plateau, and a northern plain (surrounded by mountains), which levels off via sandy hills to the Mediterranean coastline. The coastal strip constituted a "land-bridge" between Asia and Africa from time immemorial. Egyptian Pharaohs used it to invade Canaan and Phoe-

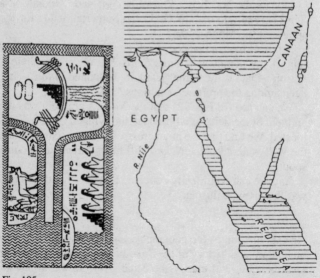

Fig. 105

nicia and to challenge the Hittites. Sargon of Akkad claimed that he reached and "washed his weapons" in the Mediterranean; "the sea lands"—the lands along the Mediterranean coast—"three times I encircled; Tilmun my hand captured." Sargon II, king of Assyria in the eighth century B.C., asserted that he had conquered the area stretching "from Bit-Yahkin on the shore of the Salt Sea as far as the border of Tilmun." The name "Salt Sea" has survived to this day as a Hebrew name for the Dead Sea—another confirmation that Tilmun lay in proximity to the Dead Sea.

Several Assyrian kings mention the Brook of Egypt as a geographic landmark on their expeditions to Egypt. Sargon II lists the Brook after describing the conquest of Ashdod, the Philistine city on the Mediterranean coast. Esarhaddon, who ruled somewhat later, boasted thus: "I trod upon Arza at the Brook of Egypt; I put Asuhili, its king, in fetters. . . . Upon Qanayah, king of Tilmun, I imposed tribute." The name "Brook of Egypt" is identical to the biblical name for the large and extensive Sinai *wadi* (shallow river that runs with water only during the rainy season) now called *Wadi El-Arish*. Ashurbanipal, who followed Esarhaddon on the throne of Assyria, claimed that he "laid his yoke of over-lordship from Tyre, which is in the Upper Sea (Mediterranean) as far as Tilmun which is in the Lower Sea" (the Red Sea).

In all instances, the geography and topography of Tilmun fully match the Sinai peninsula.

Except for annual variations, the peninsula's climate in historical times is believed to have been the same as nowadays: an irregular rainy season lasting from October through May; the rest of the year is completely dry. The meager rainfall qualifies the whole of Sinai to be defined as a desert (less than ten inches of rainfall per annum). Yet the high granite peaks in the south are snowbound in winter, and in the northern coastal strip the water level is only a few feet below the ground.

Typical to most of the peninsula are the *wadis*. In the

south, the waters of the swift and short rainfalls drain off either eastward (to the Gulf of Elat) or (mostly) westward, into the Gulf of Suez. It is there that most of the picturesque deep canyon-like wadis with flourishing oases are found. But the bulk of the peninsula's rainwater is drained northward into the Mediterranean Sea, via the extensive Wadi El-Arish and its myriad tributaries, that look on a map as the blood vessels of a giant heart. In this part of the Sinai, the depths of the wadis that make up this network may change from a few inches to a few feet; the width—from a few feet to a mile and more after a sizeable rain.

Even in the rainy season the rainfall pattern is totally erratic. Sudden downpours alternate with long dry spells. An assumption of plentiful water during the season or in its immediate aftermath could thus be very misleading. This must have happened to the Israelites as they left Egypt in mid-April and entered the Sinai Wilderness a few weeks later. Finding themselves without the expected waters, it required the intervention of the Lord twice, to show Moses where to strike the rocks for water.

The *Bedouin* (local nomads), as other seasoned travelers in the Sinai, can duplicate the miracle, if the soil making up the wadi's bed is right. The secret is that in many places the rocky bed lies above a layer of clay soil that captures the water as it quickly seeps through the rocks. With knowledge and luck, a little digging in a completely dry wadi bed uncovers water only a few feet below the surface.

Was this nomad art the great miracle performed by the Lord? Recent discoveries in the Sinai throw a new light on the subject. Israeli hydrologists (associated with the Weizmann Institute of Science) have discovered that, like parts of the Sahara Desert and some deserts in Nubia, there is "fossil water"—the remains of prehistoric lakes from another geological era—deep under the central Sinai. The vast underground reservoir, with enough water (they estimated) to suffice for a population as large as Israel's for almost one hundred years, extends for some 6,000 square miles in a wide belt that begins near the Suez Canal and reaches under Israel's arid Negev.

Though lying on the average some 3,000 feet below the rocky ground, the water is sub-artesian and rises by its own pressure to about 1,000 feet below ground. Egyptian experimental drillings for oil in the center of the northern plain (at Nakhl), have struck instead this water reservoir. Other drillings confirmed this incredible fact: above ground—an arid wilderness; below, within easy reach of modern drilling and pumping equipment—a lake of pure, sparkling water!

Could the Nefilim, with their space-age technology, have missed this knowledge? Was this, rather than a little water in a dry wadi bed, the water that gushed forth after Moses had struck the rock, as indicated by the Lord? Take in thy hand the staff with which you performed the miracles in Egypt, the Lord told Moses; you will see me standing upon a certain rock; strike that rock with the staff, "and there shall come out of it water, and the people shall drink"—enough water for a multitude of people and their livestock. So that the greatness of Yahweh be known, Moses was to take with him to the site some witnesses; and the miracle took place "before the eyes of the elders of Israel."

A Sumerian tale concerning Tilmun relates an almost identical event. It is a tale of bad times caused by a shortage of water. Crops withered, cattle were not fed, animals went thirsty, the people fell silent. Ninsikilla, spouse of Tilmun's ruler Enshag, complained to her father Enki:

> The city which thou hast given . . .
> Tilmun, the city thou hast given . . .
> Has not waters of the river . . .
> Unbathed is the maiden;
> No sparkling water is poured in the city.

Studying the problem, Enki concluded that the only solution would be to *bring up subterranean waters*. The depths must have been greater than what could be attained by digging a usual well. So Enki conceived a plan whereby the layers of rocks would be penetrated by *a missile fired from the skies!*

> Father Enki answered Ninsikilla, his daughter:
> "Let divine Utu position himself in the skies.
> A missile let him tightly affix to his 'breast'
>   and from high direct it toward the earth . . .
> From the source whence issues Earth's waters,
>   let him bring thee sweet water from the earth."

So instructed, Utu/Shamash proceeded to bring up water from the subterranean sources:

> Utu, positioning himself in the skies,
>   a missile tightly tied to his "breast,"
> From high directed it toward the earth . . .
> He let go of his missile from high in the sky.
> Through the crystal stones he brought up water;
> From the source whence issues Earth's waters
>   he brought her sweet water, from the earth.

Could a missile shot from the skies pierce the earth and cause potable water to come up? Anticipating the incredulity of his readers, the ancient scribe affirmed at the tale's end: "Verily, it was so." The miracle, the text went on, did work: Tilmun became a land "of crop raising fields and farms which bear grain"; and Tilmun-City "became port city of the Land, the site of quays and mooring piers."

The parallels between Tilmun and Sinai are thus doubly affirmed: first, the existence of the subterranean water reservoir, below the rocky surface; secondly, the presence of Utu/Shamash (the Spaceport's commander) in the proximity.

The Sinai peninsula can also account for *all* the products for which Tilmun was renowned.

Tilmun was a source of gemstones akin to the bluish lapis

lazuli which the Sumerians cherished. It is an established fact that the Pharaohs of Egypt obtained the blue-green gemstone turquoise as well as a blue-green mineral (malachite) from the southwestern parts of the Sinai. The earliest turquoise mining area is now called Wadi Magharah—the Wadi of Caves; there, tunnels were cut into the rocky sides of the wadi's canyon and miners went in to chisel out the turquoise. Later on, mining also took place at a site now named Serabit-el-Khadim. Egyptian inscriptions dating back to the Third Dynasty (2700–2600 B.C.) have been found at Wadi Magharah, and it is believed that it was then that the Egyptians began to station garrisons and occupy the mines on a continuing basis.

Archaeological discoveries, as well as depictions by the first Pharaohs of defeated and captured "Asiatic Nomads" (Fig. 106), convince scholars that at first the Egyptians only raided mines developed earlier by Semitic tribesmen. Indeed, the Egyptian name for turquoise, *mafka-t* (after which they called the Sinai the "Land of Mafkat"), stems from the Semitic verb meaning "to mine, to extract by cutting." These mining areas were in the domain of the goddess Hathor, who

Fig. 106

was known both as "Lady of Sinai" and "Lady of *Mafkat.*"
A great goddess of olden times, one of the early sky gods of
the Egyptians, she was nicknamed by them "The Cow" and
was depicted with cow's horns (see Figs. 7 and 106). Her
name, *Hat-Hor,* spelled hieroglyphically by drawing a falcon
within an enclosure 🦅 , has been interpreted by scholars to
mean "House of Horus" (Horus having been depicted as a
falcon). But it literally meant "Falcon House," which affirms
our conclusions regarding the location and functions of the
Land of the Missiles.

According to the *Encyclopaedia Britannica,* "turquoise
was obtained from the Sinai peninsula before the fourth millenium B.C. in one of the world's first important hard-rock
mining operations." At that time, the Sumerian civilization
was only beginning to stir, and the Egyptian one was almost
a millenium away. Who then could have organized the mining activities? The Egyptians said it was Thoth, the god of
sciences.

In this and in the assignment of the Sinai to Hathor, the
Egyptians emulated Sumerian traditions. According to Sumerian texts, the god who organized the mining operations
of the Anunnaki was Enki, the God of Knowledge; and Tilmun, the texts attested, was allotted in pre-Diluvial times to
Ninhursag, sister of Enki and Enlil. In her youth, she was
a smashing beauty and the chief nurse of the Nefilim. But
in her old age, she was nicknamed "The Cow" and, as the
Goddess of the Date Palm, was depicted with cow's horns
(Fig. 107). The similarities between her and Hathor, and the
analogies between their domains, are too obvious to require
elaboration.

The Sinai was also a major source of copper, and the
evidence here is that the Egyptians relied mostly on raid-
ing expeditions to obtain it. To do this, they had to penetrate
deeper into the peninsula; a Pharaoh of the Twelfth Dynasty
(the time of Abraham) left us these comments of his deeds:
"Reaching the boundaries of the foreign lands with his feet;
exploring the mysterious valleys, reaching the limits of the

Fig. 107

unknown." He boasted that his men lost not a single case of the seized booty.

Recent explorations in the Sinai by Israeli scientists found ample evidence showing that "during the times of the Early Kingdom of Egypt, in the third millenium B.C., Sinai was densely inhabited by Semitic copper-smelting and turquoise-mining tribes, who resisted the penetration of Pharaonic expeditions into their territory (Beno Rothenberg, *Sinai Explorations 1967–1972*). "We could establish the existence of a fairly large industrial metallurgical enterprise. . . . There are copper mines, miners' camps and copper smelting installations, spread from the western parts of southern Sinai to as far east as Elat at the head of the Gulf of Aqaba."

Elat, known in Old Testament times as Etzion-Gaber, was indeed a "Pittsburgh of the Ancient World." Some twenty years ago, Nelson Glueck uncovered at Timna, just north of Elat, King Solomon's copper mines. The ores were taken to Etzion-Gaber, where they were smelted and refined in "one of the largest, if not *the* largest, of metallurgical centers in existence" in ancient times *(Rivers in the Desert).*

The archaeological evidence once again ties in with biblical and Mesopotamian texts. Esarhaddon, king of Assyria, boasted that "upon *Qanayah,* king of Tilmun, I imposed tribute." The *Qenites* are mentioned in the Old Testament as inhabitants of the southern Sinai, and their name literally meant "smiths, metallurgists." The tribe into which Moses married when he escaped from Egypt into the Sinai was that of the Qenites. R. J. Forbes (*The Evolution of the Smith*) pointed out that the biblical term *Qain* ("smith") stemmed from the Sumerian KIN ("fashioner").

Pharaoh Ramses III, who reigned in the century following the Exodus, recorded his invasion of these coppersmiths' dwellings and the plundering of the metallurgical center of Timna-Elat:

> I destroyed the people of Seir, of the Tribes of the *Shasu;* I plundered their tents, their people's possessions, their cattle likewise, without number. They were pinioned and brought as captive, as tribute of Egypt. I gave them to the gods, as slaves into their temples.
>
> I sent forth my men to the Ancient Country, to the great copper mines which are in that place. Their galleys carried them; others on a land-journey were upon their asses. It has not been heard before, since the reign of the Pharaohs began.
>
> The mines were found abounding in copper; it was loaded by ten-thousands into the galleys. They were sent forward to Egypt and arrived safely. It was carried and made into a heap under the palace balcony, in many bars of copper, a hundred thousand, being of the color of gold of three refinings.
>
> I allowed all the people to see them, like wonders.

It was to spend the rest of his life in the mines of Tilmun that the gods had sentenced Enkidu; and so it was that Gilgamesh conceived the plan to charter a "Ship of Egypt" and take his comrade along—since the Land of Mines and the "Land of Missiles" were both parts of the same land. Our identification matches the ancient data.

*   *   *

Before we continue with our reconstruction of historic and pre-historic events, it is important to buttress our conclusion that *Tilmun* was the Sumerian name for the Sinai peninsula. This is not what scholars have held until now; and we should analyze their contrary views, and show why they have been wrong.

A persistent school of thought, one of whose early advocates was P. B. Cornwall *(On the Location of Tilmun),* identifies Til-mun (sometimes transcribed "Dilmun") as the island of Bah-rein in the Persian Gulf. This view relies most heavily on the inscription by Sargon II of Assyria, wherein he asserted that among the kings paying him tribute was "Uperi, king of Dilmun, whose abode is situated like a fish, thirty double-hours away, in the midst of the sea where the sun rises." This statement is taken to mean that Tilmun was an island; and the scholars who hold this view identify the "Sea where the sun rises" as the Persian Gulf. They then end up with Bahrein as the answer.

There are several flaws in this interpretation. First, it could well be that only the capital city of Tilmun was on an off-shore island: the texts leave no doubt that there was a land Tilmun and a Tilmun-city. Secondly, other Assyrian inscriptions which describe cities as being "in the midst of the sea" apply to coastal cities on a bay or a promontory, but not on an island (as, for instance, Arvad on the Mediterranean coast). Then, if the "sea where the sun rises" indicates a sea east of Mesopotamia, the Persian Gulf does not qualify, since it lies to the south, not to the east, of Mesopotamia. Also, Bahrein lies too close to Mesopotamia to account for thirty double-hours of sailing. It is situated some 300 miles south of the Mesopo-tamian Gulf ports; in sixty hours of sailing, even at a leisurely pace, a distance many times greater could be covered.

Another major problem arising from a Bahrein-Tilmun identification concerns the products for which Tilmun was renowned. Even in the days of Gilgamesh, not all of the Land Tilmun was a restricted area. There was a part, as we have seen, where sentenced men toiled in dark and dusty mines, digging out the copper and gemstones for which Tilmun was famous. Long associated with Sumer in culture and trade,

Tilmun supplied it with certain desired species of woods. And its agricultural areas—subject of the above-mentioned tale of Ninsikilla's plea for artesian waters—provided the ancient world with highly prized onions and dates.

Bahrein had none of these, except for some "ordinary dates." So, to circumvent the problem, the pro-Bahrein school has developed a complex answer. Geoffrey Bibby *(Looking for Dilmun)* and others of like mind suggest that Bahrein was a trans-shipment point. The products, they agree, indeed came from some other, more distant land. But the ships which carried these goods did not go all the way to Sumer. They stopped and unloaded their goods at Bahrein, where the famous merchants of Sumer picked them up for the final haul into Sumerian ports; so that, when the Sumerian scribes wrote down where the goods had come from (so this theory goes), they wrote down "Dilmun," meaning Bahrein.

But why would ships that have sailed great distances fail to sail the final short distance to the actual destination in Mesopotamia, and instead go to the extra trouble and cost of offloading at Bahrein? Also, this theory stands in direct contradiction to specific statements by rulers of Sumer and Akkad that the ships *of* Tilmun, among ships from other lands, anchored at their port cities. Ur-Nanshe, a king of Lagash some two centuries after Gilgamesh was king of neighboring Uruk, claimed that "the ships of Tilmun . . . brought me wood as tribute." We recognize the name *Tilmun* in his inscription (Fig. 108) by the pictograph for "missile." Sargon, the first ruler of Akkad, boasted that "at the wharf of Akkad he made moor ships from Meluhha, ships from Magan and ships from Tilmun."

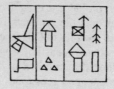

Fig. 108

Clearly, then, the ships brought the products of Tilmun straight to the Mesopotamian ports proper, as logic and economics would dictate. Likewise, the ancient texts speak of direct exports from Mesopotamia to Tilmun. One inscription records a shipment of wheat, cheese and shelled barley from Lagash to Tilmun (circa 2500 B.C.); no trans-shipment at an island is ever mentioned.

One of the leading opponents of the Bahrein theory, Samuel N. Kramer *(Dilmun, the "Land of the Living")* stressed the fact that the Mesopotamian texts described it as "a distant land," reachable not without risk and adventure. These descriptions do not match a close-by island, reachable after an easy sailing down the quiet waters of the Persian Gulf. He also attached great importance to the fact that the various Mesopotamian texts placed Tilmun near *two* bodies of water, rather than near or in a single sea. The Akkadian texts located Tilmun *ina pi narati*—"at the mouth of the two flowing waters": where two bodies of water begin.

Guided by yet another statement, which said that Tilmun was the land "where the Sun rises," Kramer concluded, first, that Tilmun was a land and not an island; and secondly, that it must have been located east of Sumer, for it is in the east that the Sun rises. Searching in the east for a place where two bodies of water meet, he could come up only with a southeastern point, where the Persian Gulf meets the Indian Ocean. Baluchistan, or somewhere near the Indus River, were his suggestions.

Kramer's own hesitation stemmed from the well-known fact that numerous Sumerian and Akkadian texts listing countries and peoples do not mention Tilmun in association with such eastern lands as Elam or Aratta. Instead, they lump together as lands situated next to each other *Meluhha* (Nubia/Ethiopia), *Magan* (Egypt) and *Tilmun.* The proximity between Egypt (Magan) and Tilmun is spelled out at the end of the "Enki and Ninhursag" text, where the appointment of Nintulla as "Lord of Magan"

and of Enshag as "Lord of Tilmun" obtains the blessing
of the two gods. It is also evident from a remarkable text,
written as an autobiography of Enki, which describes his
activities after the Deluge, assisting Mankind to establish
its civilizations; again, Tilmun is listed next to Magan and
Meluhha:

> The lands of *Magan* and *Tilmun*
>   looked up at me.
> I, Enki, moored the Tilmun-boat at the coast,
> Loaded the Magan-boat sky high.
> The joyous boat of *Meluhha*
>   transports gold and silver.

In view of this proximity of Tilmun to Egypt, what about
the statements that Tilmun was "where the sun rises"—
meaning (scholars say) *east* of Sumer, and not west of it (as
the Sinai is)?

The simple answer is that the texts do not make that statement at all. They do not say "where the *Sun* rises"; they state
"where *Shamash* ascends"—and that makes all the difference. Tilmun was not at all in the east; but it certainly was
the place where Utu/Shamash (the god whose celestial symbol was the Sun, and not the Sun itself) ascended skyward in
his rocketships. The words of the Gilgamesh epic are quite
clear:

> At the Mountain of *Mashu* he arrived,
> Where by day the *Shems* he watched
>   as they depart and come in . . .
> Rocket-men guard its gate . . .
>   they watch over *Shamash*
>   as he ascends and descends.

That, indeed was the place whereto Ziusudra had been
taken:

In the Land of the Crossing
in mountainous Tilmun—
the place *where Shamash ascends*—
they caused him to dwell.

And so it was that Gilgamesh—denied permission to mount a Shem, and seeking therefore only to converse with his ancestor Ziusudra—set his steps to *Mount Mashu* in Tilmun— the *Mount of Moshe* (Moses) in the Sinai peninsula.

Modern botanists have been amazed by the variety of the peninsula's flora, finding more than a thousand species of plants, many unique to the Sinai, varying from tall trees to tiny shrubs. Where there is water—as in oases, or below the surface in the coastal sand dunes, or in the beds of the wadis—these trees and shrubs grow with impressive persistence, having adapted themselves to the particular climate and hydrography of the Sinai.

The Sinai's northeastern parts could well have been the source of the craved-for onions. Our name for the variety with the long green stem, *scallion,* bears evidence to the port from which this delicacy was shipped to Europe: *Ascalon* on the Mediterranean coast, just north of the Brook of Egypt.

One of the trees that adapted itself to the Sinai's unique circumstances is the acacia, which accommodates its high transpiration rate by growing only in the wadi beds, where it exploits the subsurface moisture down to many feet. As a result, the tree can live for almost ten years without rain. It is a tree whose timber is a prized wood; according to the Old Testament, the Holy Ark and other components of the Tabernacle were made of this wood. It could have well served as the prized wood which the kings of Sumer imported for their temples.

An ever-present sight in the Sinai are the tamarisks—

bush-like trees that trace the wadi courses year round, for their roots reach down to the subsurface moisture and they can grow even where the water is saline and brackish. After especially rainy winters, the tamarisk groves fill up with a sweet, granular white substance which is the excretion of small insects that live on the tamarisks. The Bedouin call it by its biblical name, *manna,* to this very day.

The tree with which Tilmun was mostly associated in antiquity, however, was the *date palm.* It is still the Sinai's most important tree economically. Needing minimal cultivation, it provides the Bedouin with fruit (dates); its pulp and kernels are fed to camels and goats; the trunk is used for building and as fuel; the branches for roofing; the fibers for rope and weaving.

We know from Mesopotamian records that these dates were also exported from Tilmun in antiquity. The dates were so large and tasty that recipes for the meals of the gods of Uruk (the city of Gilgamesh) specified that "every day of the year, for the four daily meals, 108 measures of ordinary dates, and dates of the Land Tilmun, as also figs and raisins . . . shall be offered to the deities." The nearest and most ancient town on the land route from Sinai to Mesopotamia was Jericho. Its biblical epithet was "Jericho, the city of dates."

The date palm, we find, has been adopted as a symbol in Near Eastern religions, i.e., in ancient concepts of Man and his gods. The biblical Psalmist promised that "the Righteous like a date palm shall flourish." The Prophet Ezekiel, in his vision of the rebuilt temple of Jerusalem, saw it decorated with alternating "Cherubim and date palms . . . so that a date palm was between a Cherub and a Cherub, and two (date palms) flank each Cherub." Residing at the time among the exiles whom the Babylonians had forcefully brought over from Judea, Ezekiel was well acquainted with the Mesopotamian depictions of the Cherubim and date palm theme (Fig. 109).

Alongside the Winged Disk (the emblem of the Twelfth Planet), the symbol most widely depicted by all the ancient nations was the *Tree of Life.* Writing in *Der Alte Orient,* Felix

Fig. 109

von Luschau has shown back in 1912 that the Greek Ionian column capitals (Fig. ll0a) as well as Egyptian ones (Fig. ll0b) were in fact stylizations of the Tree of Life in the shape of a date palm (Fig. ll0c), and confirmed earlier suggestions that the Fruit of Life of legend and epic tales was some special species of the date fruit. We find the theme of the date palm as the symbol of Life carried on even in Muslim Egypt, as in the decorations of Cairo's grand mosque (Fig. ll0d).

Many major studies, such as *De Boom des Levens en Schrift en Historie* by Henrik Bergema and *The King and the Tree of Life in Ancient Near Eastern Religion* by Geo. Widengren, show that the concept of such a tree, growing in an Abode of the Gods, has spread from the Near East all over Earth and has become a tenet of all religions, everywhere.

The source of all these depictions and beliefs were the Sumerian records of the Land of the Living,

> Tilmun,
> Where old woman says not "I am an old woman,"
> Where old man says not "I am an old man."

The Sumerians, masters of word-plays, called the Land of the Missiles TIL.MUN; yet the term could also mean "Land Of Living," for TIL also meant "Life." The Tree of Life in Sumerian was GISH.TIL; but GISH also meant a man-made, a manufactured object; so that GISH.TIL could also mean "The Vehicle to Life"—a rocketship. In art too, we find

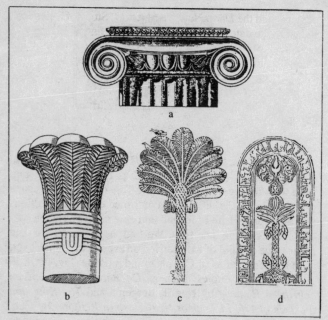

Fig. 110

the Eagle-men sometimes saluting not the date palm, but a rocket (Fig. 60).

The binding knots tighten further, as we find that in Greek religious art, the omphalos was associated with the date palm. An ancient Greek depiction of Delphi shows that the omphalos replica that was erected outside Apollo's temple was set up next to a date palm (Fig. 111). Since no such trees grow in Greece, it was an artificial tree made (scholars believe) of bronze. The association of the omphalos with the date palm must have been a matter of basic symbolism, for these depictions were repeated also in respect to other Greek oracle centers.

We have found earlier that the omphalos served as a link between Greek, Egyptian, Nubian and Canaanite "oracle centers" and the *Duat*. Now we find this "Stone of Splendor" linked to the date palm—the Tree of the Land of Living.

Indeed, Sumerian texts accompanying depictions of the Cherubim included the following incantation:

> The dark-brown tree of Enki I hold in my hand;
> The tree that tells the count, great heavensward weapon,
> I hold in my hand;
> The palm tree, great tree of oracles, I hold in my hand.

A Mesopotamian depiction shows a god holding up in his hand this "palm tree, Great tree of oracles" (Fig. 112). He is

Fig. 111

Fig. 112

granting this Fruit of Life to a king at the place of the "Four Gods." We have already come upon this place, in Egyptian texts and depictions: they were the Four Gods of the Four Cardinal Points, located by the Stairway to Heaven in the *Duat*. We have also seen (Fig. 72) that the Sumerian Gateway to Heaven was marked by the date palm.

And we have no more doubt that the target of the ancient Search for Immortality was a Spaceport—somewhere in the Sinai peninsula.

XI

The Elusive Mount

Somewhere in the Sinai peninsula, the Nefilim had established their post-Diluvial Spaceport. Somewhere in the Sinai peninsula, mortals—a select few, with their god's blessing—could approach a certain mountain. There, "Go back!" the guarding bird-men ordered Alexander, "for the land on which you stand belongs to God, alone." There, "Do not come nearer!" the Lord called out to Moses, "for the place whereon thou standest is sacred ground." There, eagle-men challenged Gilgamesh with their stun-rays, only to realize he was no mere mortal.

The Sumerians called this mount of encounter Mount MA.SHU—the Mount of the Supreme Barge. The tales of Alexander named it Mount *Mushas*—the Mountain of *Moses*. Its identical nature and function, coupled with its identical name, suggest that in all instances it was the same mountain that was the destination's landmark. It thus seems that the answer to the question "Where in the peninsula was the gateway?" is right at hand: Is not the Mount of the Exodus, "Mount Sinai," clearly marked on maps of the peninsula—the tallest peak among the high granite mountains of southern Sinai?

The Israelite Exodus from Egypt has been commemorated each year for the past thirty-three centuries by the celebration of the Passover. The historical and religious records of the Hebrews are replete with references to the Exodus, the wanderings in the Wilderness, the Covenant at Mount Sinai. The people have been constantly reminded of the Theophany, when the whole nation of Israel had seen the Lord Yahweh alight in his glory upon the sacred mount. Yet its location was de-emphasized, lest attempts be made to make the place a cult center. There is no recorded instance in the Bible of anyone even trying to pay a return visit to Mount Sinai, with one exception: the Prophet Elijah. Some four centuries after the Exodus, he escaped for his life after having slain the priests of Ba'al upon Mount Carmel. Setting his course to the mount in Sinai, he lost his way in the desert. An angel of the Lord revived him and placed him in a cave in the mount.

Nowadays, it would seem, one needs no guiding angel to find Mount Sinai. The modern pilgrim, as pilgrims have done for centuries past, sets his course to the monastery of Santa Katarina (Fig. 113), so named after the martyred Katherine of Egypt whose body angels carried to the nearby peak bearing her name. After an overnight stay, at daybreak, the pilgrims begin the climb to *Gebel Mussa* ("Mount Moses" in Arabic). It is the southern peak of a two mile massif rising south of the monastery—the "traditional" Mount Sinai with which the Theophany and the Lawgiving are associated (Fig. 114).

The climb to that peak is long and difficult, involving an ascent of some 2,500 feet. One path is by way of some 4,000 steps laid out by the monks along the western slopes of the massif. An easier way that takes several hours longer begins in the valley between the massif and a mountain appropriately named after Jethro, the father-in-law of Moses, and rises gradually along the eastern slopes until it connects with the last 750 steps of the first path. It was at that intersection, according to the monk's tradition, that Elijah encountered the Lord.

A Christian chapel and a Muslim shrine, both small and crudely built, mark the spot where the Tablets of the Law were given to Moses. A cave nearby is revered as the "cleft in the rock" wherein the Lord placed Moses as He passed by him, as related in Exodus 33:22. A well along the descent

Fig. 113

Fig. 114

route is identified as the well from which Moses watered the flock of his father-in-law. Every possible event relating to the Holy Mount is thus assigned by the monks' traditions a definite spot on the peak of Gebel Mussa and its surroundings.

From the peak of Gebel Mussa, one can see some of the other peaks which make up the granite heartland, of which this mount is a member. Surprisingly, it appears to be lower than many of its neighbors!

Indeed, in support of the Saint Katherine legend, the monks have put up a sign in the main building which proclaims:

Altitude	5012 FT
Moses Mount	7560 FT
Sta. Katherine Mount	8576 FT

As one is convinced that Mount Katherine is indeed the higher one—in fact, the highest in the peninsula—and thus rightly chosen by the angels to hide the saint's body thereon, one is also disappointed that—contrary to long-held beliefs—God had brought the Children of Israel to this forbidding area, to impress upon them his might and his laws not from the tallest mount around.

Had God missed the right mountain?

In 1809, the Swiss scholar Johann Ludwig Burckhardt arrived in the Near East in behalf of the British Association for Promoting the Discovery of the Interior Parts of Africa. Studying Arabic and Muslim customs, he put a *turban* on his head, dressed as an Arab and changed his name to Ibrahim Ibn Abd Allah—Abraham the Son of Allah's Servant. He was thus able to travel in parts hitherto forbidden to the infidels, discovering ancient Egyptian temples at Abu Simbel and the Nabatean rock city of Petra in Transjordan.

On April 15, 1816, he set out on camelback from the town

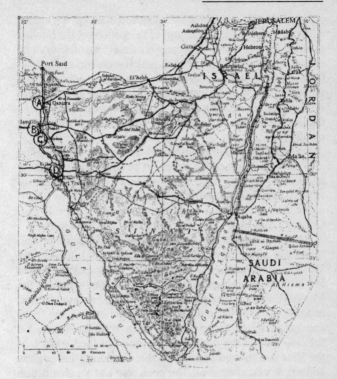

Fig. 115

of Suez, at the head of the Gulf of Suez. His goal was to
retrace the route of the Exodus, and thereby to establish the
true identity of Mount Sinai. Following the presumed route
taken by the Israelites, he traveled south along the western
coast of the peninsula. There the mountains begin some ten
to twenty miles away from the coast, creating a desolate
coastal plain cut here and there by wadis and a couple of hot
springs, including one favored by the Pharaohs.

As he went south, Burckhardt noted the geography, topography, distances. He compared conditions and place names with the descriptions and names of the stations of the Exodus as mentioned in the Bible. Where the limestone plateau ends, nature has provided a sandy belt which separates the plateau from a belt of Nubian sandstone, serving as a cross-Sinai avenue. There Burckhardt turned inland, and after a while set his course southward into the granite heartland, reaching the Katherine monastery from the north (as today's air traveler does).

Some of his observations are of a lingering interest. The area, he found, produced excellent dates; the monks had a tradition of sending large boxes of them as an annual tribute to the sultan in Constantinople. Befriending the area's Bedouins, they invited him to the annual feast in honor of "St. George"; they called him "El Khidher"—The Evergreen!

Burckhardt ascended mounts Mussa and Katherine and toured the area extensively. He was especially fascinated by *Mount Umm Shumar*—a mere 180 feet shorter than Mount St. Katherine—which rises somewhat southwest of the Mussa-Katherine group. From a distance, its top dazzled in the sun "with the most brilliant white color," due to an unusual inclusion of particles of mica in the granite rocks, forming "a striking contrast with the blackened surface of the slate and the red granite" of the mountain's lower parts and the surrounding area. The peak also had the distinction of offering an unobstructed view to both the Gulf of Suez ("el-Tor was distinctly visible") and the Gulf of Aqaba (Gulf of Elat). Burckhardt found it mentioned in the convent's records, that Umm Shumar used to be a principal location of monastic settlements. In the fifteenth century, "caravans of asses laden with corn and other provisions passed by this place regularly from the convent to el-Tor, for this is the nearest road to that harbor."

His way back was via Wadi Feiran and its oasis—the largest in Sinai. Where the wadi leaves the mountains and reaches the coastal strip, Burckhardt climbed up a magnificent mountain rising over 6,800 feet—*Mount Serbal,* one of the tallest in the peninsula. There he found remains of shrines and pilgrims' inscriptions. Additional research established that the main monastic center in Sinai, through most of the centuries, was at Wadi Feiran, near Serbal—and not at St. Katherine.

When Burckhardt published his findings *(Travels in Syria and the Holy Land)*, his conclusions shook the scholarly and biblical world. The true Mount Sinai, he stated, was not Mount Mussa, but Mount Serbal!

Inspired by Burckhardt's writings, the French Count Léon de Laborde toured the Sinai in 1826 and 1828; his main contribution to the knowledge of the area *(Commentaire sur L'Exode)* were his fine maps and drawings. He was followed in 1839 by the Scottish artist David Roberts; his magnificent drawings, wherein he embellished accuracy with some imaginative flair, aroused great interest in an era before photography.

The next major journey to Sinai was undertaken by the American Edward Robinson, together with Eli Smith. Like Burckhardt, they left Suez City on camelback, armed with his book and de Laborde's maps. It took them thirteen early spring days to reach St. Katherine. There, Robinson gave the monks' legends a thoroughgoing examination. He found out that at Feiran there indeed was a superior monastic community, sometimes led by full bishops, to which Katherine and several other monastic communities in southern Sinai were subordinate; so that tradition must have placed greater emphasis on Feiran. In the tales and documents, he discovered that mounts Mussa and Katherine were of no Christian consequence in the early Christian centuries, and that Katherine's supremacy developed only in the seventeenth century, when the other unfortified monastic communities fell prey to invaders and marauders. Checking local Arab traditions, he found that the biblical names "Sinai" and "Horeb" were totally unknown to the local Bedouins; it was the Katherine monks who began to apply these names to certain mountains.

Was Burckhardt, then, right? Robinson *(Biblical Re-searches in Palestine, Mount Sinai and Arabia Petraea)* found a problem with the route by which Burckhardt had the Israelites reach Serbal, and therefore refrained from endorsing the new idea; but he shared the doubts regarding Mount Mussa, and pointed at another nearby mountain as a better choice.

The possibility that the long-held tradition identifying Mount Sinai with Mount Mussa was incorrect was a challenge that the great Egyptologist and founder of scientific archaeology, Karl Richard Lepsius, could not resist. He crossed the Gulf of Suez by boat, landing at *el-Tor* ("The Bull")—the harbor town where Christian pilgrims to St. Katherine and Mount Moses used to land even before the Muslims made it a major stopover and decontamination center on the sea route from Egypt to Mecca. Nearby rose the majestic Mount Umm-Shumar, which Lepsius on and off compared as a "candidate" with Mussa and Serbal. But after extensive research and area touring, he focused on the burning problem of that day: Mussa or Serbal?

His findings were published in *Discoveries in Egypt, Ethiopia and the Peninsula of Sinai 1842–1845* and *Letters from Egypt, Ethiopia and Sinai,* the latter including (in translation from German), the full text of his reports to the king of Prussia, under whose patronage he traveled. Lepsius voiced doubts regarding Mount Mussa almost as soon as he reached the area: "The remoteness of that district, its distance from frequented roads of communication and its position in the lofty range," he wrote, ". . . rendered it peculiarly applicable for individual hermits; but for the same reason inapplicable for a large people." He felt certain that the hundreds of thousands of Israelites could not have subsisted among the desolate granite peaks of Mount Mussa for the long (almost a year) Israelite stay at Mount Sinai. The monastic traditions, he confirmed, dated to the sixth century A.D. at the earliest; they could therefore serve as no guide in this quest.

Mount Sinai, he stressed, was in a desert plain; it was also called in the Scriptures Mount *Horeb,* the Mount of the Dryness. Mussa was amidst other mountains and not in a desert area. On the other hand, the coastal plain in front of Mount Serbal was such an area—large enough to hold the Israelite multitudes as they viewed the Theophany; and the adjoining Wadi Feiran was the only place that could sustain them and their cattle for a year. Moreover, only possession of "this unique fertile valley" could have justified the Amalekite attack (at Rephidim, a gateway place near Mount Sinai); there was no such fertile place, worth fighting for, near Mount Mussa. Moses first came to the mount in search of grazing for his flock; this he could find at Feiran, but not at the desolate Mount Mussa.

But if not Mount Mussa, why Mount Serbal? Besides its "correct" location at Wadi Feiran, Lepsius found some concrete evidence. Describing the mount in glowing terms, he reported finding on its top "A deep mountain hollow, around which the five summits of Serbal unite in a half circle and form a towering crown." In the middle of this hollow he found ruins of an old convent. It was at that hallowed spot, he suggested, that the "Glory of the Lord" had landed, in full view of the Israelites (who were gathered in the plain to the west). As to the fault that Robinson had found with Burckhardt's Exodus route to Serbal—Lepsius offered an alternative detour which corrected the problem.

When the conclusions of the prestigious Lepsius were published, they shook tradition in two ways: he emphatically denied the identification of Mount Sinai with Mount Mussa, voting for Serbal; and he challenged the Exodus route previously taken for granted.

The debate that followed raged for almost a quarter of a century and produced discourses by other researchers, notably Charles Foster *(The Historical Geography of Arabia; Israel in the Wilderness)* and William H. Bartlett *(Forty Days in the Desert on the Track of the Israelites).* They added suggestions, confirmations and doubts. In 1868 the British government joined the Palestine Exploration Fund in sending a full-scale expedition to Sinai. Its mission, in addition to extensive geodesic and mapping work, was to establish once and for all the route of the Exodus and the location of Mount Sinai. The group was led by captains Charles W. Wilson and Henry Spencer Palmer of the Royal Engineers; it included Professor Edward Henry Palmer, a noted Orientalist and Arabist. The expedition's official report *(Ordnance Survey of the Peninsula of Sinai)* was enlarged upon by the two Palmers, in separate works.

Previous researchers went to the Sinai for brief tours mostly in springtime. The Wilson-Palmer expedition departed from Suez on November 11, 1868, and returned to Egypt on April 24, 1869—staying in the peninsula from the beginning of winter until the following spring. Thus, one of its first discoveries was that the mountainous south gets very cold in winter and that it snows there, making passage difficult, if not impossible. The higher peaks, such as Mussa and Katherine, remain snowcovered for many winter months. The Israelites—who had never seen snow in Egypt—had stayed a year in this area. Yet there is no mention at all in the Bible of either snow or even cold weather.

While Captain Palmer *(Sinai: Ancient History from the Monuments)* provided data on the archaeological and historical evidence uncovered (early habitations, Egyptian presence, inscriptions in the first known alphabet), it was the task of Professor E. H. Palmer *(The Desert of the Exodus)* to outline the group's conclusions regarding the route and the mount.

In spite of lingering doubts, the group vetoed Serbal and voted for the Mount Mussa location, but with a twist. Since in front of Mount Mussa there was no valley wide enough where the Israelites could encamp and see the Theophany, Palmer offered a solution: The correct Mount Sinai was not the southern peak of the massif (Gebel Mussa), but its northern peak, *Ras-Sufsafeh,* which faces "the spacious plain of Er-Rahah where no less than two million Israelites could encamp." In spite of the long-held tradition, he concluded, "we are compelled to reject" Gebel Mussa as the Mount of the Lawgiving.

The views of Professor Palmer were soon criticized, supported or modified by other scholars. Before long, there were several southern peaks that were offered as the true Mount Sinai, as well as several different routes to choose from.

But was the southern Sinai the only place in which to search?

Back in April 1860, the *Journal of Sacred Literature* published a revolutionary suggestion, that the Holy Mount was not in southern Sinai at all, but should be looked for in the central plateau. The anonymous contributor pointed out that its name, *Badiyeth el-Tih,* was very significant: it meant "the Wilderness of the Wandering," and the local Bedouins explain that it was there that the Children of Israel wandered. The article suggested a certain peak of the *el-Tih* as the proper Mount Sinai.

So, in 1873, a geographer and linguist named Charles T. Beke (who explored and mapped the origins of the Nile) set out "in search of the *true* Mount Sinai." His research established that Mount Mussa was so named after a fourth century monk Mussa who was famed for his piety and miracles, and not after the biblical Moses; and that the claims for Mount Mussa were begun only circa A.D. 550. He also pointed out that the Jewish historian Josephus Flavius (who recorded his people's history for the Romans after the fall of Jerusalem in A.D. 70) described Mount Sinai as the highest in its area, which ruled out both Mussa and Serbal.

Beke also asked, how could the Israelites have gone south at all, past the Egyptian garrisons in the mining areas? His question has remained one of the unanswered objections to a southern location of Mount Sinai.

Charles Beke will not be remembered as the man who finally found the true Mount Sinai: as the title of his work indicated *(Discoveries of Sinai in Arabia and Midian),* he concluded that the mount was a volcano, somewhere southeast of the Dead Sea. But he raised many questions which cleared the desk for fresh and unfettered thinking regarding the location of the mount and the route of the Exodus.

The search for Mount Sinai in the southern part of the peninsula was closely linked with the notion of the "Southern Crossing" and "Southern Route" of the Exodus. These held that the Children of Israel literally crossed the Red Sea (from west to east) at or through the head of the Gulf of Suez. Once across, they were out of Egypt and on the western shores of the Sinai peninsula. They then journeyed south along the coastal strip, turned (somewhere) inland, and reached Mount Sinai (as, say, Burckhardt had done).

The Southern Crossing was indeed a deep-rooted and plausible tradition, buttressed by several legends. According to Greek sources, Alexander the Great was told that the Israelites had crossed the Red Sea at the head of the Gulf of Suez; it was there that he tried to emulate the Crossing.

The next great conqueror known to have attempted the feat was Napoleon, in 1799. His engineers established that where the head of the Gulf of Suez sends a "tongue" inland, south of where Suez City is located, there exists an underwater ridge, some 600 feet wide, which extends from coast to coast. Daredevil natives cross there at ebb tide, with the waters up to their shoulders. And if a strong east wind blows, the seabed is almost cleared of all water.

Napoleon's engineers worked out for their emperor the right place and time for emulating the Children of Israel. But an unexpected change in the wind's direction brought a sudden onrush of waters, covering the ridge with more than seven feet of water within minutes. The great Napoleon escaped with his life in the nick of time.

These experiences only served to convince nineteenth century scholars that it was indeed at that end of the Gulf of Suez that the miraculous Crossing had taken place: a wind could create a dry path, and a change in wind could indeed sink an army soon thereafter. On the opposite, Sinai, side of the Gulf, there was a place named *Gebel Murr* ("The Bitter Mountain") and near it *Bir Murr* ("The Bitter Well"), invitingly fitting as Marah, the place of bitter waters, encountered by the Israelites after the Crossing. Further south lay the oasis of *Ayun Mussa*—"The Spring of Moses"; now was not this the next station, Elim, remembered for its beautiful springs and numerous date palms? The Southern Crossing thus seemed to fit well with the Southern Route theory, no matter where the turn inland had taken place further on.

The Southern Crossing also agreed with the then current notions regarding Egypt in antiquity and the Israelite bondage therein. Egypt's historical heart was the Heliopolis-Memphis hub, and it was assumed that the Israelites slaved in the construction of the nearby pyramids of Gizeh. From there, a route led almost straight east, toward the head of the Gulf of Suez and the Sinai peninsula beyond it.

But as archaeological discoveries began to fill in the historical picture and provide an accurate chronology, it was established that the great pyramids were built some fifteen centuries before the Exodus—more than a thousand years before the Hebrews even came to Egypt. The Israelites, more and more scholars agreed, must have toiled in the construction of a new capital which the Pharaoh Ramses II had built circa 1260 B.C. It was named *Tanis* and it was located in the northeastern part of the Delta. The Israelite abode—the land of Goshen—was consequently presumed to have been in the northeast rather near the center of Egypt.

The construction of the Suez Canal (1859–1869), which was accompanied by the accumulation of topographical, geological, climatic and other data, confirmed the existence of a natural rift which in an earlier geological age may have joined the Mediterranean Sea in the north with the Gulf of Suez in the south. That link had shrunk for various reasons, leaving behind a watery chain consisting of the marshy lagoons of Lake Manzaleh, the small lakes Ballah and Timsah, and the joined Great and Little Bitter Lakes. All these lakes may have been larger at the time of the Exodus, when the head of the Gulf of Suez probably extended farther inland.

Archaeological work complementing the engineering data also established that there existed in antiquity two "Suez Canals," one connecting Egypt's hub with the Mediterranean and the other to the Gulf of Suez. Following natural wadi beds or dried up branches of the Nile, they carried "sweet" water for drinking and irrigation and were navigable. The finds confirmed that in earlier times there was indeed an almost continuous water barrier which served as Egypt's eastern border.

The engineers of the Suez Canal prepared in 1867 the following diagram (Fig. 116) of a north-south section of the Isthmus, identifying four ridges of high ground which must have served in antiquity, as they still do, the gateways to and from Egypt through the watery barrier (Fig. 115):

(A) Between the marshy lagoons of Manzaleh and Lake Ballah—the modern crossing town of *el-Qantara* ("The Span").

(B) Between Lake Ballah and Lake Timsah—the modern crossing point of *Ismailiya*.

(C) Between Lake Timsah and the Great Bitter Lake—a ridge known in Greek-Roman times as the *Serapeum*.

(D) Between the Little Bitter Lake and the head of the Gulf of Suez—a "land-bridge" known as The *Shalouf*.

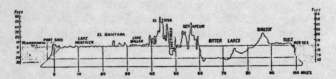

Fig. 116

Through these Gateways, a number of Routes connected Egypt with Asia via the Sinai peninsula. One has to bear in mind that the crossing of the Red Sea (or Sea/Lake of Reeds) was not premeditated: it took place only after the Pharaoh changed his mind about letting the Israelites go; whereupon the Lord commanded them to turn back from the edge of the desert which they had already reached, and "encamp by the sea." Therefore, they originally exited from Egypt by one of the usual gateways; but which one?

DeLesseps, the Canal's master builder, voiced the opinion that they used Gateway "C," south of Lake Timsah. Others, like Olivier Ritter *(Histoire de l'Isthme de Suez),* concluded from the exact same data that it was Gateway "D." In 1874, the Egyptologist Heinrich Karl Brugsch, addressing the International Congress of Orientalists, identified the landmarks connected with the Israelite enslavement and Exodus in the northeastern corner of Egypt. Therefore, he said, the logical gateway was all the way north—Gateway "A."

As it turned out, such a theory of a Northern Crossing was nearly a century old when Brugsch launched it, having been suggested in *Hamelneld's Biblical Geography* back in 1796, and by various researchers thereafter. But Brugsch, as even his adversaries conceded, presented the theory with a "really brilliant and dazzling array of claimed corroboratory evidence from the Egyptian monuments." His paper was published the following year under the title *L'Exode et les Monuments Egyptiens.*

In 1883, Edouard H. Naville *(The Store City of Pithom and the Route of the Exodus)* identified *Pithom,* the Israelite slave-labor city, at a site west of Lake Timsah. This and other identifications and evidence offered by others (such as by George Ebers in *Durch Gosen zum Sinai*) established that the Israelite abode extended from Lake Timsah westward, not northward. Goshen was not in the extreme northeast of Egypt, but adjoined the center of the watery barrier.

H. Clay Trumbull *(Kadesh Barnea)* then offered what has since been generally accepted as the correct identification for

Succoth, the starting point of the Exodus: it was a common caravan gathering place west of Lake Timsah, and Gateway "B" was the nearest at hand. But it was not taken, as stated in Exodus 13:17-18: "And it came to pass, when Pharaoh let the people go, that the Lord did not lead them the Way of the Land of the Philistines, though it was near . . . and the Lord turned the people by the Way of the Desert *Yam Suff.*" Thus, Trumbull suggested, the Israelites ended up at Gateway "D"; pursued by the Pharaoh, they crossed through the waters of the head of the Gulf of Suez.

As the nineteenth century drew to a close, scholars raced to give the final word on the subject. The views of the "southerners" were emphatically summed up by Samuel C. Bartlett *(The Veracity of the Hexateuch):* the Crossing was in the south, the Route led south, Mount Sinai was in the south of the peninsula *(Ras-Sufsafeh).* With equal decisiveness, such scholars as Rudolf Kittel *(Geschichte der Hebräer),* Julius Wellhausen *(Israel und Judah),* and Anton Jerku *(Geschichte des Volkes Israel)* offered the opinion that the Northern Crossing meant a *northern* Mount Sinai.

One of their strongest arguments (now generally accepted by scholars) was that *Kadesh-Barnea,* where the Israelites sojourned for most of their forty years in the peninsula, was not a chance station but a premeditated target of the Exodus. It has been firmly identified as the fertile area of the Ain-Kadeis ("Spring of Kadesh") and Ain-Qudeirat oases in northeastern Sinai. According to Deuteronomy 1:2, Kadesh-Barnea was situated "eleven days" from Mount Sinai. Kittel, Jerku and others of like opinion therefore selected mountains in the vicinity of Kadesh-Barnea as the true Mount Sinai.

In the last year of the nineteenth century, H. Holzinger *(Exodus)* offered a compromise: the Crossing was at "C"; the Route led south. But the Israelites turned inland well before reaching the Egyptian-garrisoned mining areas. Their route led via the highland plateau of the *el-Tih,* the "Wilderness of the Wandering." They then circled northward through the flat Central Plain, toward a Mount Sinai *in the north.*

*   *   *

As the twentieth century began, the focus of research and debate shifted to the question: What was the *route* of the Exodus?

The ancient coastal route, which the Romans called *Via Maris*—"Way of the Sea"—began at el-Qantara ("A" on map). Though it led through shifting sand dunes, it was blessed with water wells all along its course, and the date palms amazingly growing out of the barren sands provide sweet fruit in season and welcome shade all year round.

The second route, beginning at Ismailiya ("B"), runs almost parallel to the coastal road but some twenty to thirty miles south of it, through undulating hills and occasional low mountains. The natural wells are sparse, and the subterranean water level lies deep below the sand and sandstones: artificial wells must be dug several hundred feet to reach water. A traveler—even nowadays, even by car (the paved highways follow the ancient paths)—soon realizes that he is in a real desert.

From earliest times, the Way of the Sea was preferred by armies that had naval support; the more inland route—harsher though it was—was taken by those who sought to be safe from (or unseen by) the naval and coastal patrols.

Gateway "C" could lead either to Route "B," or to the twin routes which extended from Gateway "D" through a mountain chain into the Sinai's Central Plain. The hard, flat ground of the Central Plain does not allow deep wadi beds. During the winter rains, some wadis overfill and give the appearance of small lakes—lakes in the desert! The waters soon flow off, but some filter down through the gravel and clay that make up the wadi beds; it is there that digging can literally bring water out of the ground.

The more northerly route extending from Gateway "D" led the traveler via the Giddi Pass, past the northern mountainous rim of the Central Plain, on to Beersheba, Hebron and Jerusalem. The more southerly route, via the Mitla Pass, bears the Arabic name *Darb el Hajj*—"Way of the Pilgrims." This route was the early way for Muslim pilgrims from Egypt to the holy city of Mecca in Arabia. Starting near Suez City, they crossed a desert strip and went through the mountains via the Mitla Pass; then journeyed across the Central Plain to the oasis of *Nakhl* (Fig. 117) where a fort, pilgrims' inns and water pools had been built. From there they continued southeast to reach Aqaba at the head of the Gulf of Aqaba, whence they moved along the Arabian coast to Mecca.

Which of these four possible routes—the "Ways" of the Bible—had the Israelites taken?

In the aftermath of the Northern Crossing presentation by Brugsch, much was made of the biblical statement regarding the "Way of the Land of the Philistines" which was not taken, "though it was near." The Bible continued the statement with the following explanation: "For the Lord said: 'Lest the people repent when they see war, and return to Egypt.'" It has been assumed that this "Way of the Land of the Philistines" was the coastal route (which began at gateway "A"), the way the Pharaohs preferred for their military and trade expeditions, and which was strung with Egyptian forts and garrisons.

At the turn of the century, A. E. Haynes, a captain in the Royal Engineers, studied Sinai's routes and water resources under the auspices of the Palestine Exploration Fund. In his

Fig. 117

published report on "The Route of the Exodus" he revealed impressive familiarity not only with biblical scriptures, but also with the work of previous researchers, including the Rev. F. W. Holland (who visited the Sinai five times) and Major-General Sir C. Warren (who paid particular attention to water supplies in the "Wilderness of the Wandering" of the Central Plain).

Captain Haynes focused on the problem of the Route-That-Was-Not-Taken. Unless it was a handy and obvious way for reaching the Israelite's goals—why was it mentioned at all as a viable alternative? He pointed out that Kadesh-Barnea—by then accepted as a premeditated goal of the Exodus—indeed lay within easy reach of the coastal route. Therefore, he concluded. Mount Sinai, situated on the way to Kadesh, also had to be located within easy reach of the coastal route, whether or not this route was finally taken.

Barred from the coastal Route "A," Captain Haynes concluded, it was "the probable plan of Moses" to lead the Israelites directly to Kadesh, with a stop at Mount Sinai, via Route "B." But the Egyptian pursuit and the Crossing of the Red Sea may have forced a detour via routes "C" or "D." The Central Plain was indeed the "Wilderness of Wandering." *Nakhl* was an important station near Mount Sinai, before or after reaching it. The mount itself had to be located about 100 miles from Kadesh-Barnea, which equals (Captain Haynes estimated) the biblical distance of "eleven days." His candidate was Mount *Yiallaq,* a limestone mountain "of most impressive dimensions, lying like a huge barnacle" on the northern rim of the Central Plain—"exactly halfway between Ismailiyah and Kadesh." Its name, which he spelled *Yalek,* "approximates closely to the ancient *Amalek,* the prefix *Am* meaning 'country of.'"

In the years that followed, the possibility of an Israelite journey via the Central Plain gained supporters; some (as Raymond Weill, *Le Séjour des Israélites au désert du Sinai*) accepted the Mount-near-Kadesh theory; others (as Hugo Gressmann, *Mose und seine Zeit*) believed that the Israelites turned from Nakhl not northeast but southeast, toward Aqaba. Others—Black, Bühl, Cheyne, Dillmann, Gardiner, Grätz, Guthe, Meyer, Musil, Petrie, Sayce, Stade—agreed or disagreed partly or completely. As all the scriptural and geographical arguments were exhausted, it seemed that only an actual field test could resolve the issue. But how does one duplicate the Exodus?

World War I (1914–1918) was the answer, for the Sinai soon became the arena of a major struggle between the British on the one hand and the Turks and their German allies on the other hand. The prize of these campaigns was the Suez Canal.

The Turks lost no time in crossing into the Sinai peninsula, and the British quickly withdrew from their main administrative-military centers at El-Arish and Nakhl. Unable to advance by the desirable "Way of the Sea," for the same old reason that the Mediterranean was controlled by the enemy's (British) navy, the Turks amassed a herd of 20,000 camels to carry water and supplies for an advance on the Canal via route "B" to Ismailiyah. In his memoirs, the Turkish Commander, Djemal Pasha *(Memories of a Turkish Statesman, 1913–1919)* explained that "the great problem, on which everything hangs in these difficult military operations in the desert of Sinai, is the question of water. In any other than the rainy season it would be impossible to cross this waste with an expeditionary force of approximately 25,000 men." His attack was repulsed.

The German allies of the Turks then took matters in hand. For their motorized equipment, they preferred the hard, flat Central Plain for an advance on the Canal. With the aid of water engineers, they discovered the subterranean resources and dug a network of wells all along their lines of communication and advance. Their attack in 1916 also failed. When the British took the offensive in 1917, they naturally advanced along the coastal route. They reached the old demarcation line at Rafah in February 1917; within months they captured Jerusalem.

The British memoirs on the Sinai fighting by General A. P. Wavell *(The Palestine Campaigns)* has a bearing on our subject primarily by his admission that the British High Command estimated that their enemies could not find in the Central Plain water for more than 5,000 men and 2,500 camels. The German side of the Sinai campaigns is told in *Sinai* by Theodor Wiegand and the commanding general, F. Kress von Kressenstein. The military endeavor is described against the background of terrain, climate, water sources and history, coupled with an impressive familiarity with all previous research. Not surprisingly, the conclusions of the German military men parallel the conclusions of the British military men: no marching columns, no multitudes of men and beasts could be led through the southern granite mountains. Devoting a special chapter to the question of the Exodus, Wiegand and von Kressenstein asserted that "the region of Gebel Mussa cannot come into consideration for the biblical Mount Sinai." They were of the opinion that it was "the monumental Gebel *Yallek*"—echoing the conclusions of Captain Haynes. Or, they added, perhaps as Guthe and other German scholars have suggested, Gebel *Maghara,* which rises opposite Gebel Yallek, on the northern side of Route "B."

One of Britain's own military men, who was governor of the Sinai after World War I, became acquainted with the peninsula during his long tenure there as perhaps no single person in modern times until then. Writing in *Yesterday and Today in Sinai,* C. S. Jarvis too asserted that there was no way the Israelite multitudes (even if their numbers were smaller than 600,000, as W.M.F. Petrie had suggested) and their livestock could have traveled through—much less sustained themselves for more than a year—in the "tumbled mass of pure granite" of the southern Sinai.

To the known arguments, he added new ones. It had already been suggested that the *manna* which served in lieu of bread was the edible, white, berry-like resinous deposit left by small insects that feed on the tamarisk bushes. There are few tamarisks in southern Sinai; they are plentiful in north-

ern Sinai. Next fact concerned the quails, which provided the
meat to eat. These birds migrate from their native southern
Russia, Rumania and Hungary to winter in the Sudan (south
of Egypt); they return northward in the spring. To this day,
the Bedouins easily catch the tired birds as they alight on
the Mediterranean coast after long flights. The quails do not
come to the southern Sinai; and if they did, they could not
possibly fly over the high peaks of that area.

The whole drama of the Exodus, Jarvis insisted, was played
out in the northern Sinai. The "Sea of Reeds" was the Serbonic Sealet (*Sebkhet el Bardawil* in Arabic) from which the
Israelites marched south-southeast. Mount Sinai was Gebel
Hallal—"a most imposing limestone massif over 2,000 feet
high and standing in the midst of a vast alluvial plain all
by itself." The mountain's Arabic name, he explained, meant
"The Lawful"—as befits the Mount of the Lawgiving.

In the years that followed, the most pertinent research on
the subject was conducted by scholars of the Hebrew University of Jerusalem and other Hebrew institutions of higher
learning in what was then Palestine. Combining their inti-
mate knowledge of the Hebrew Bible and other scriptures
with thorough on-site investigations in the peninsula, few
found support for the southern location tradition.

Haim Bar-Deroma *(Hanagev* and *Vze Gvul Ha'aretz)*
accepted a Northern Passage but believed that the Route
then took the Israelites south, through the Central Plain, to
a volcanic Mount Sinai in Transjordan. Three noted scholars—F. A. Theilhaber, J. Szapiro and Benjamin Maisler *(The
Graphic Historical Atlas of Palestine: Israel in Biblical
Times)*—accepted the Northern Passage via the shoal of the
Serbonic Sea. El-Arish, they said, was the verdant oasis of
Elim; Mount Hallal was Mount Sinai. Benjamin Mazar, in
various writings and in *Atlas Litkufat Hatanach,* adopted the
same position. Zev Vilnay, a biblical scholar who hiked in
Palestine and Sinai literally from end to end *(Ha'aretz Bami-
kra),* opted for the same route and mount. Yohanan Aharoni
(The Land of Israel in Biblical Times), accepting the possi-

bility of a Northern Passage, believed that the Israelites journeyed toward Nakhl in the Central Plain; but then proceeded
to a Mount Sinai in the south.

As the debate continued to engross the scholarly and biblical world, it became apparent that the basic unresolved
issue was this: Insofar as the Crossing was concerned, the
weight of the evidence negated a northern body of water;
but insofar as Mount Sinai was concerned, the weight of the
evidence negated a southern location. The impasse focused
the attention of scholars and explorers on the only remaining compromise: the Central Plain of the Sinai peninsula. In
the 1940s, M. D. Cassuto *(Commentary on the Book of Exodus* and other writings) facilitated acceptance of the central
route idea by showing that the Route-Not-Taken ("The Way
of the Land of the Philistines") was not the long-held sea
route, but the more inland route "B." Therefore, a Crossing
via Gateway "C" leading southeast to the Central Plain was
in full accord with the biblical narrative—without requiring
a continued journey to the south of the peninsula.

The long occupation of the Sinai by Israel, in the aftermath of the 1967 war with Egypt, opened up the peninsula
to study and research on an unprecedented scale. Archaeologists, historians, geographers, topographers, geologists,
engineers examined the peninsula from tip to toe. Of particular interest have been the explorations by the teams of
Beno Rothenberg *(Sinai Explorations 1967–1972* and other
reports), mostly under the auspices of Tel-Aviv University.
In the northern coastal strip, many ancient sites reflected
the "bridge-like nature of this area." In the Central Plain of
north Sinai, no ancient sites of permanent abode were found,
but only evidence of camping sites, attesting that this was
only a transit area. When the camping sites were plotted on
the map, they formed "a clear line from the Negev toward
Egypt, and this should be considered as the direction of prehistoric movements across the 'Desert of the Wanderings'
(the el-Tih)."

It was against this newly understood background of the

ancient Sinai that a Hebrew University biblical geographer, Menashe Har-El, offered a new theory *(Massa'ei Sinai).* Reviewing all the arguments, he pointed out the submerged ridge (see Fig. 116) which rises between the Great and the Little Bitter Lakes. It is shallow enough to be crossed if a wind blows away the waters; it was there that the Crossing had taken place. Then the Israelites followed the traditional route south; passing Marrah *(Bir Murrah)* and Elim *(Ayun Mussa),* they reached the shores of the Red Sea and encamped there.

Here Har-El offered his major innovation: having journeyed along the Gulf of Suez, the Israelites did not go all the way south. They proceeded only some twenty miles to the mouth of *Wadi Sudr*—and followed the wadi's valley into the Central Plain, proceeding via Nakhl to Kadesh-Barnea. Har-El identified Mount Sinai with Mount *Sinn-Bishr* which rises some 1,900 feet at the entrance to the wadi, and suggested that the battle with the Amalekites had actually taken place on the coast of the Gulf of Suez. This suggestion has been rejected by Israeli military experts familiar with the terrain and history of warfare in the Sinai.

Where, then, was Mount Sinai? We must look again at the ancient evidence.

The Pharaoh, in his Journey to the Afterlife, went eastward. Crossing the watery barrier, he set his course to a pass in the mountains. He then reached the *Duat,* which was an oval-shaped valley surrounded by mountains. The "Mountain of Light" was situated where the Stream of Osiris divided into tributaries.

The pictorial depictions (Fig. 16) showed the Stream of Osiris meandering its way through an agricultural area, distinguished by its ploughmen.

We have found similar pictorial evidence from Assyria. The Assyrian kings, it should be remembered, arrived at the Sinai from the opposite direction to that of the Egyptian kings: from the northeast, via Canaan. One of them, Esarhaddon, engraved on a stela what amounts to a route map of his own quest for "Life" (Fig. 118). It shows the date palm—the code emblem for the Sinai; a farming area symbolized by the plough; and a "Sacred Mount." In the upper register we see Esarhaddon at the shrine of the Supreme Deity, near the Tree of Life. It is flanked by the sign of the bull—the very same image (the "golden calf") that the Israelites had fashioned at the foot of Mount Sinai.

All this does not bespeak the harsh, barren granite peaks of southern Sinai. Rather, it suggests northern Sinai and its dominant *Wadi El-Arish,* whose very name means Stream of the Husbandman. It is among its tributaries, in a valley surrounded by mountains, that the Mount was located.

There is only one such place in the whole of the Sinai peninsula. Geography, topography, historical texts, pictorial depictions—all point at the *Central Plain* in Sinai's northern half.

Even E. H. Palmer, who went so far as to invent the Ras-Sufsafeh twist in order to uphold the southern identification,

Fig. 118

knew in his heart that a desert that stretches as far as the eye can see, and not a peak in a sea of granite mountains, was the location of the Theophany and the wanderings of the Israelites.

"The popular conception of Sinai," he wrote in *The Desert of the Exodus,* "even in the present day, seems to be that a single isolated mountain which may be approached from any direction rises conspicuously above a boundless plain of sand. The Bible itself, if we read it without the light of modern discovery, certainly favors this idea. . . . Mount Sinai is always alluded to in the Bible as though it stood alone and unmistakable in the midst of a level desert plain."

There indeed exists such a "level desert plain" in the Sinai peninsula, he admitted; but it is not covered with sand: "Even in those parts [of the peninsula] which approach most nearly to our conception of what a desert ought to be—a solid ocean bounded only by the horizon or by a barrier of distant hills—sand is the exception, and the soil resembles rather a hard gravel path than a soft and yielding beach."

He was describing the Central Plain. To him, the absence of sand marred the "desert" image; to us, its hard gravel top meant that it was admirably suited for the Spaceport of the Nefilim. And if Mount Mashu marked the gateway to the Spaceport, it had to be located on the outskirts of this facility.

Have then generations of pilgrims gone south in vain? Did the veneration of the southern peaks begin only with Christianity? The discovery by archaeologists atop these mounts of shrines, altars, and other evidence of worship from olden days attests differently; and the many inscriptions and rock carvings (including the Jewish Candelabra emblem) by pilgrims from many faiths and over many millenia bespeak a veneration going back to Man's earliest acquaintance with the area.

As one almost wishes there were *two* "Mounts Sinai" to

satisfy both tradition and facts, it turns out that such notions too are not new. Even before the last two centuries of concerted effort to identify the Mount, biblical and theological scholars had wondered whether the various biblical names for the Sacred Mount did not indicate that there originally were two sacred mountains, not one. These names included "Mount Sinai" (the Mountain of/in Sinai), which was the Mount of the Lawgiving; "Mount Horeb" (the Mountain of/in the Dryness); "Mount Paran," which was listed in Deuteronomy as the mount in Sinai from which Yahweh had appeared unto the Israelites; and "the Mountain of the Gods," where the Lord first revealed himself unto Moses.

The geographic location associated with two of the names is decipherable. Paran was the wilderness adjoining Kadesh-Barnea, possibly the biblical name for the Central Plain; so that "Mount Paran" had to be located there. It was to that Mount that the Israelites had gone. But the Mount where Moses had his first encounter with the Lord, "the Mountain of the Gods," could not have been too far from the Land of Midian; for "Moses was shepherding the flock of Jetro, his father-in-law, the priest of Midian; and he led the flock unto the wilderness, and came unto the Mountain of the Gods, unto Horeb." The abode of the Midianites was in southern Sinai, along the Gulf of Aqaba and astride the copper-working areas. "The Mountain of the Gods" must have been located somewhere in an adjoining wilderness—in southern Sinai.

There have been found Sumerian cylinder seals depicting the appearance of a deity unto a shepherd. They show the god appearing from between two mountains (Fig. 119), with a rocket-like tree behind him—perhaps the *Sneh* ("Burning Bush") of the biblical tale. The introduction of two peaks in the shepherd scene fits the frequent biblical reference to the Lord as *El Shaddai*—God of the Two Peaks. It thus raises yet another distinction between the Mount of the Lawgiving and the Mountain of the Gods: the one was a solitary mount in a desert plain; the other seems to have been a combination of two sacred peaks.

Fig. 119

The Ugaritic texts too recognize a "Mountain of the young gods" in the environs of Kadesh, and two peaks of El and Asherah—*Shad Elim, Shad Asherath u Rahim*—in the south of the peninsula. It was to that area at *mebokh naharam* ("Where the two bodies of water begin"), *kerev apheq tehomtam* ("Near the cleft of the two seas") that El had retired in his old age. The texts, we believe, describe the southern tip of the Sinai peninsula.

There was, we conclude, a Gateway Mount on the perimeter of the Spaceport in the Central Plain. And there were two peaks in the peninsula's southern tip that also played a role in the comings and goings of the Nefilim. They were the two peaks that *measured up*.

XII

The Pyramids of Gods and Kings

Somewhere in the vaults of the British Museum there is stashed away a clay tablet which was found at Sippar, the "cult center" of Shamash in Mesopotamia. It shows him seated on a throne, under a canopy whose pillar is shaped as a date palm (Fig. 120). A king and his son are introduced to Shamash by another deity. In front of the seated god there is mounted upon a pedestal a large emblem of a ray-emitting planet. The inscriptions invoke the gods Sin (father of Shamash), Shamash himself and his sister Ishtar.

The theme of the scene—the introduction of kings or priests to a major deity—is a familiar one, and poses no problems. What is unique and puzzling in this depiction are the two gods (almost superimposed upon one another) who, from somewhere outside of where the introduction is taking place, hold (with two pairs of hands) two cords leading to the celestial emblem.

Who are the two Divine Cordholders? What is their function? Are they identically situated, and if so, why do they hold or pull two cords, and not just one? Where are they? What is their connection with Shamash?

Sippar, scholars know, was the seat of the High Court of Sumer; Shamash was consequently the ultimate lawgiver. Hammurabi, the Babylonian king famous for his law code, depicted himself receiving the laws from an enthroned Shamash. Was the scene with the two Divine Cordholders

also somehow connected with lawgiving? In spite of all the
speculation, no one has so far come up with an answer.

The solution, we believe, has been available all along, in
the very same British Museum—not among its "Assyrian"
exhibits, however, but in its Egyptian Department. In a room
separate from the mummies and the other remains of the dead
and their tombs, there are exhibited pages from the various
papyri inscribed with the *Book of the Dead.* And the answer is
right there, for all to see (Fig. 121).

It is a page from the "Papyrus of Queen Nejmet" and the
drawing illustrates the final stage of the Pharaoh's journey
in the *Duat.* The twelve gods who pulled his barge through
the subterranean corridors have brought him to the last corridor, the Place of Ascending. There, the "Red Eye of Horus"
was waiting. Then, shed of his earthly clothing, the Pharaoh
was to ascend heavenward, his Translation spelled out by the
beetle hieroglyph ("Rebirth"). Gods standing in two groups
pray for his successful arrival at the Imperishable Star.

And, unmistakably, there in the Egyptian depiction are
two Divine Cordholders!

Without the congestion of the depiction from Sippar, this
one from the Book of the Dead shows the two Cordholders not crowding out each other, but at two different ends of
the scene. They are clearly located outside of the subterranean corridor. Moreover: each site manned by a Cordholder
is marked by an *omphalos* resting upon a platform. And, as
the action imparted by the drawing shows, the two divine
aides are not simply holding the cords, but are engaged in
measuring.

The discovery should not surprise: have not the verses of
the Book of the Dead described how the journeying Pharaoh
encounters the gods "who hold the rope in the *Duat,*" and the
gods "who hold the measuring cord"?

A clue in the *Book of Enoch* now comes to mind. There, it
will be recalled, it is related that as he was taken by an angel
to visit the earthly paradise in the west, Enoch "saw in those
days how long cords were given to angels who took to them-

Fig. 120

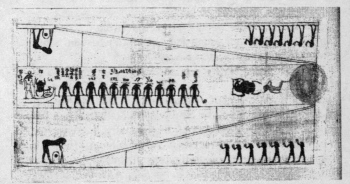

Fig. 121

selves wings, and they went towards the north." In reply to
Enoch's question, his guiding angel explained: "They have
gone off to measure . . . they shall bring the measures of the
Righteous to the Righteous . . . all these measures shall reveal the secrets of the Earth."

Winged beings going north to measure. . . . Measures
that shall reveal the secrets of the Earth. . . . All at once,
the words of the Prophet Habakuk thunder in our ears—the
words describing the appearance of the Lord from the south,
going north:

> The Lord from South shall come,
> The Holy One from Mount Paran.
> Covered are the heavens with his halo,
> His splendor fills the Earth;
> His brilliance is like light.
> His rays shine forth
>   from where his power is concealed.
> The Word goes before him,
>   sparks emanate from below.
> He pauses to measure the Earth;
> He is seen, and the nations tremble.

Was the measuring of Earth and its "secrets" then related
to the powered flight of the gods in Earth's skies? The Ugaritic texts add a clue as they tell us that, from the peak of
Zaphon, Ba'al "a cord strong and supple stretches out, heavenwards (and) to the Seat of Kadesh."

Whenever these texts report a message from one god to
another, the verse begins with the word *Hut*. Scholars assume that it was a kind of a calling prefix, a kind of "Are you
ready to hear me?" But the term could literally mean in the
Semitic languages "cord, rope." Significantly, the term *Hut*
in Egyptian also means "to extend, to stretch out." Heinrich
Brugsch, commenting on an Egyptian text dealing with the
battles of Horus *(Die Sage von der geflügten Sonnenscheibe)*
pointed out that Hut was also a place name—the abode of

the Winged Extenders, as well as the name of the mountain within which Horus was imprisoned by Seth.

We find in the Egyptian depiction (Fig. 121) that the conical "oracle stones" were located where the Divine Measurers were stationed. Baalbek too was the location of such an omphalos, a Stone of Splendor that could perform the *Hut* functions. There was an oracle stone at Heliopolis, the Egyptian twin-city of Baalbek. Baalbek was the gods' Landing Platform; the Egyptian cords led to the Pharaoh's Place of Ascent in the *Duat*. The biblical Lord—called in Habakuk by a variant of *El*—measured Earth as he flew from south to north. Are these all just a series of coincidences—or parts of the same jigsaw puzzle?

And then we have the depiction from Sippar. It is not puzzling if we recall that in pre-Diluvial times, when Sumer was the Land of the Gods, Sippar was the Spaceport of the Anunnaki, and Shamash its Commander. Thus viewed, the role of the Divine Measurers will become clear: *their cords measured out the path to the Spaceport.*

It would help to recall how Sippar was established, how the site of the first Spaceport on Earth was determined, some 400,000 years ago.

When Enlil and his sons were given the task of creating a Spaceport upon planet Earth, in the plain between the Two Rivers of Mesopotamia, a master plan was drawn up; it involved the selection of a site for the Spaceport, the determination of a flight path, and the establishment of guidance and Mission Control facilities. Based on the most conspicuous natural feature in the Near East—*Mount Ararat*—a north-south meridian was drawn through it. A flight path over the Persian Gulf, well away from flanking mountain ranges, was marked out at the precise and easy angle of 45°. Where the two lines intersected, on the banks of the Euphrates River, Sippar—"Bird City"—was to be.

Five settlements, equidistant from one another, were laid out along the diagonal 45° line. The central one—Nippur ("The Place of Crossing")—was to serve as the Mission

Control Center. Other settlements marked out an arrow-like corridor; all the lines converged at Sippar (Fig. 122).

All that, however, was wiped out by the Deluge. In its immediate aftermath—some 13,000 years ago—only the Landing Platform at Baalbek had remained. Until a replacement Spaceport could be built, all landings and takeoffs of the Shuttlecraft had to be conducted there. Are we to assume that the Anunnaki relied on reaching the site, tucked away between two mountain ranges, by sheer skilled piloting—or

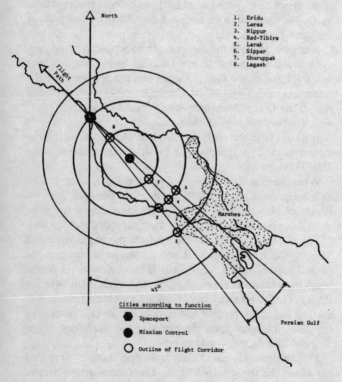

Fig. 122

can we safely surmise that as soon as possible they worked out an arrow-like Landing Corridor to Baalbek?

With the aid of photographs of Earth from spacecraft of the U.S. National Aeronautics and Space Administration, we can view the Near East as the Anunnaki had seen it from their own craft (Fig. 123). There, a dot in the north, was Baalbek. What vantage points could they choose from which to mark out a triangular landing corridor? Close at hand, to the southeast, rose the granite peaks of southern Sinai. Amid the granite core rose the highest peak (now called Mount St. Katherine). It could serve as a natural beacon to outline the southeastern line. But where was the counterpoint in the northwest, on which the northern line of the triangle could be anchored?

Aboard the Shuttlecraft, the Surveyor—a "Divine Measurer"—glanced at the earthly panorama below, then studied his maps again. In the far distance, beyond Baalbek, there loomed the twin-peaked Ararat. He drew a straight line from Ararat through Baalbek, extending it all the way into Egypt.

He took his compass. With Baalbek as the focal point, he

Fig. 123

drew an arc through the highest peak of the Sinai peninsula. Where it intersected the Ararat-Baalbek line, he made a cross within a circle. Then he drew two lines of equal length, one connecting Baalbek with the peak in Sinai, the other with the site marked by the cross (Fig. 124).

This, he said, will be our triangular Landing Corridor, to lead us straight to Baalbek.

But sir, one of those aboard said, there is nothing there, where you have made the cross—nothing that can serve as a guiding beacon!

We will have to erect there a *pyramid,* the commander said.

And they flew on, to report their decision.

Had such a conversation indeed taken place aboard a shuttle-craft of the Anunnaki? We, of course, shall never know (un-

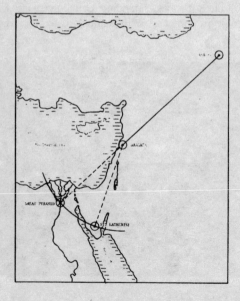

Fig. 124

less a tablet is someday found recording the event); we have merely dramatized some astounding but *undeniable* facts:

- The unique platform at Baalbek has been there from bygone days, and it is still there intact in its enigmatic immensity;
- Mount St. Katherine is still there, rising as the highest peak of the Sinai peninsula, hallowed since ancient days, enveloped (together with its twin-peaked neighbor, Mount Mussa) in legends of gods and angels;
- The Great Pyramid of Giza, with its two companions and the unique Sphinx, is situated precisely on the extended Ararat-Baalbek line; and
- The distance from Baalbek to Mount St. Katherine and to the Great Pyramid of Giza is exactly the same.

This, let us add at once, is only part of the amazing grid which—as we shall show—was laid out by the Anunnaki in connection with their post-Diluvial Spaceport. Therefore, whether or not the conversation had taken place aboard a shuttlecraft, we are pretty certain that *that is how the pyramids came to be in Egypt.*

There are many pyramids and pyramidical structures in Egypt, dotting the landscape from where the Nile breaks out into its delta in the north, all the way south to (and into) Nubia. But when one speaks of The Pyramids, the many emulations, variations, and "mini-pyramids" of later times are omitted, and scholars and tourists alike focus on the twenty-odd pyramids believed to have been erected by Pharaohs of the Old Kingdom (circa 2700–2180 B.C.). These, in turn, consist of two distinct groups: the pyramids clearly identified with rulers of the Fifth and Sixth Dynasties (such as Unash, Teti, Pepi), which are elaborately decorated and inscribed with the renowned Pyramid Texts; and the older pyramids attributed to kings of the Third and Fourth Dynasties. It is the latter, much older and first-ever, pyramids, that are the most intriguing. Much grander, more solid, more accurate, more perfect than all those that followed them, they are also the most mysterious—for they contain not a clue to reveal the secret of their construction. Who built them, how were they built, why, even when—no one can really say; there are only theories and educated guesses.

The textbooks will tell us that the first of Egypt's imposing pyramids was built by a king named Zoser, the second Pharaoh of the Third Dynasty (circa 2650 B.C. by most counts). Selecting a site west of Memphis, on the plateau that served as the necropolis (city of the dead) of that ancient capital, he instructed his brilliant scientist and architect named Imhotep to build him a tomb that would surpass all previous tombs. Until then, the royal custom was to carve out a tomb in the rocky ground, bury the king, and then cover the grave with a giant horizontal tombstone called a *mastaba* that in time grew to substantial dimensions. The ingenious Imhotep, some scholars hold, covered the original mastaba over the tomb of Zoser with layer upon layer of ever smaller mastabas, in two phases (Fig. 125a), achieving a step pyramid. Beside it, within a large rectangular courtyard, a variety of functional and decorative buildings were erected—chapels, funerary temples, storehouses, attendants' quarters and so on; the whole area was then surrounded by a magnificent wall. The pyramid and the ruins of some of the adjoining buildings and the wall can still be seen (Fig. 125b) at Sakkara—a name believed to have honored Seker, the "hidden God."

The kings who followed Zoser, the textbooks continue to explain, liked what they saw and tried to emulate Zoser. Presumably it was Sekhemkhet, who followed Zoser on the throne, who began to build the second step pyramid, also at Sakkara. It never really got off the ground, for reasons unknown (perhaps the missing ingredient was the enigmatic genius of science and engineering, Imhotep). A third step pyramid—or rather the mound containing its ruined beginnings—was discovered about midway between Sakkara and Giza to the north. Smaller than the previous ones, it is

Fig. 125

logically attributed by some scholars to the next Pharaoh on the throne, named Khaba. Some scholars believe that there were one or two additional attempts by unidentified kings of the Third Dynasty to build pyramids here and there, but without much success.

We now have to go some thirty miles south of Sakkara, to a place named Maidum, to view the pyramid deemed to have been the next one chronologically. In the absence of evidence, it is logically presumed that this pyramid was built by the next Pharaoh in line, named Huni. Through much circumstantial evidence, it is held however that he only began the construction, and that the attempt to complete the pyramid was undertaken by his successor, Sneferu, who was the first king of the Fourth Dynasty.

It was commenced, as the previous ones, as a step pyramid. But for reasons which remain totally unknown and for which even theories are lacking, its builders decided to make it a "true" pyramid, namely to provide it with smooth sides. This meant that a smooth layer of stones was to be fitted as an outer skin at a steep angle (Fig. 126a). Again for reasons unknown, an angle of 52° was selected. But what, according to the textbooks, was to be the first-ever true pyramid ended as a dismal failure: the outer stone skin, the stone fillings and parts of the core itself collapsed under the sheer weight of the stones, all set one atop the other at a precarious angle. All that remains of that attempt is part of the solid core, with a large mound of debris all around it (Fig. 126b).

Some scholars (as Kurt Mendelssohn, *The Riddle of the Pyramids*) suggest that Sneferu was building at the same time another pyramid, somewhat north of Maidum, when the Maidum pyramid collapsed. The architects of Sneferu then hurriedly changed the pyramid's angle in mid-construction. The flatter angle (43°) assured greater stability and reduced the height and mass of the pyramid. It was a wise decision, as witness the fact that the pyramid—appropriately called the Bent Pyramid (Fig. 127)—still stands.

Encouraged by his success, Sneferu ordered another true pyramid to be built near the first. It is referred to as the Red Pyramid, due to the hue of its stones. It is supposed to have represented the realization of the impossible: a triangular shape rising from a square base; its sides measuring about 656 feet each, its height a staggering 328 feet. The triumph, however, was not achieved without a little cheating: instead of the perfect inclination of 52°, the sides of this "first classical pyramid" rise at the much safer angle of under 44°. . . .

We now arrive chronologically, scholars suggest, at the epitome of Egyptian pyramid buildings.

Sneferu was the father of Khufu (whom Greek historians called Cheops); it has thus been assumed that the son followed up the achievement of his father by building the next true pyramid—only a larger and grander one: the Great Pyramid of Giza. It stands majestically as it has stood for millenia in the company of two other major pyramids, attributed to his successors Chefra (Chephren) and Menka-ra (Mycerinus); the three are surrounded by smaller satellite pyramids,

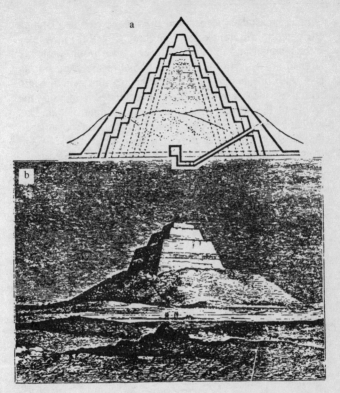

Fig. 126

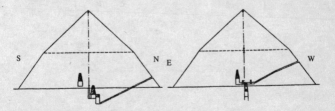

S   N   E   W

Fig. 127

temples, mastabas, tombs and the unique Sphinx. Though attributed to different rulers, the three (Fig. 128) obviously were planned and executed as a cohesive group, perfectly aligned not only to the cardinal points of the compass but also with one another. Indeed, triangulations which begin with these three monuments can be extended to measure the whole of Egypt—the whole of Earth, for that matter. This was first realized in modern times by Napoleon's engineers: they selected the apex of the Great Pyramid as the focal point from which they triangulated and mapped Lower Egypt.

This was made even easier by the discovery that the site is located, for all intents and purposes, right on the thirtieth

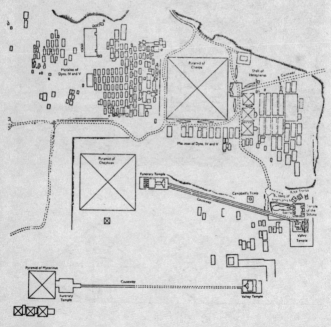

Fig. 128

parallel (north). The whole Giza complex of massive monuments had been erected at the eastern edge of the Libyan Plateau, which begins in Libya in the west and stretches to the very banks of the Nile. Though only some 150 feet above the river's valley below, the Giza site has a commanding and unobstructed view to the four horizons. The Great Pyramid stands at the extreme northeastern edge of a protrusion of the plateau; a few hundred feet to the north and east, sands and mud begin, making such massive structures impossible. One of the first scientists to have taken precise measurements, Charles Piazzi Smyth *(Our Inheritance in the Great Pyramid)* established that the center of the Great Pyramid was at northern latitude 29° 58' 55"—a mere one-sixtieth of a degree off from exactly at the thirtieth parallel. The center of the second large pyramid was only thirteen seconds ($^{13}/_{3600}$ of a degree) to the south of that.

The alignment with the cardinal points of the compass; the inclination of the sides at the perfect angle of about 52° (at which the height of the pyramid in relation to its circumference is the same as that of a radius of a circle to its circumference); the square bases, set on perfectly level platforms—all bespeak of a high degree of scientific knowledge of mathematics, astronomy, geometry, geography and of course building and architecture, as well as the administrative ability to mobilize the necessary manpower, to plan and execute such massive and long-term projects. The wonderment only increases as one realizes the *interior* complexities and precision of the galleries, corridors, chambers, shafts and openings that have been engineered within the pyramids, their hidden entrances (always on the north face), the locking and plugging systems—all unseen from the outside, all in perfect alignment with each other, all executed within these artificial mountains as they were being built layer after layer.

Though the Second Pyramid (that of Chefra) is only slightly smaller than the First, "Great Pyramid" (heights: 470 and 480 feet; sides at base 707 and 756 feet, respectively), it is the latter that has by and large captured the interest and imagination of scholars and laymen since men ever set their eyes upon these monuments. It has been and still remains the largest stone building in the world, having been constructed of an estimated 2,300,000 to 2,500,000 slabs of yellow limestone (the core), white limestone (the smooth facing or casing), and granite (for interior chambers and galleries, for roofing, etc.). Its total mass, estimated at some 93 million cubic feet weighing 7 million tons, has been calculated to exceed that of all the cathedrals, churches and chapels combined that have been built in England since the beginning of Christianity.

On ground that has been artificially leveled, the Great Pyramid rises on a thin platform whose four corners are marked by sockets of no ascertained function. In spite of the passage of millenia, continental shifts, Earth's wobble around its own axis, earthquakes and the immense weight of the pyramid itself, the relatively thin platform (less than twenty-two inches thick) is still undamaged and perfectly level: the error or shift in its perfect horizontal alignment is less than a tenth of an inch over the 758 feet that each side of the platform measures.

From a distance, the Great Pyramid and its two companions appear to be true pyramids; but when approached it is realized that they too are a kind of step pyramid, built layer upon layer (scholars call them courses) of stone, each layer smaller than the one below it. Modern studies, in fact, suggest that the Great Pyramid is a step pyramid at its core, engineered to sustain great vertical stress (Fig. 129). What gave it the smooth, inclined sides were the casing stones with which its sides were covered. These have been removed in Arab times and used for the construction of nearby Cairo; but a few can still be seen in position near the top of the Second Pyramid, and some were discovered at the base of the Great Pyramid (Fig. 130). It is these casing stones which determined the angle of the pyramid's sides; they are the heaviest of all the stones used to build the pyramid proper; the six faces that each stone has have been cut and polished to an

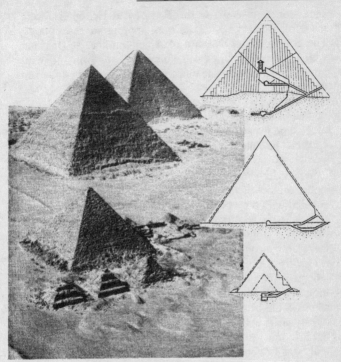

Fig. 129

accuracy of optical standards—they fitted not only the core stones which they covered, but also each other on all four sides, forming a precision-made area of twenty-one acres of limestone blocks.

The Giza pyramids are nowadays also minus their apex or capstones, which were shaped as pyramidions and may have been either made of metal or covered with a shiny metal—as the similar pyramidion-shaped tips of obelisks were. Who, when and why they were removed from their great heights, no one knows. It is known however that in later times these apex stones, resembling the *Ben-Ben* at Heliopolis, were made of special granite and bore appropriate inscriptions. The one from the pyramid of Amen-em-khet at Dahshur, which was found buried some distance away from the pyramid (Fig. 131), bore the emblem of the Winged Globe and the inscription

> The face of king Amen-em-khet is opened,
> That he may behold the Lord of the Mountain of Light
> When he sails across the sky.

When Herodotus visited Giza in the fifth century, the capstones are not mentioned, but the pyramids' sides were still covered with the smooth facings. As others before and after him, he wondered how these monuments—counted among the Seven Wonders of the ancient world—were ever built. Regarding the Great Pyramid, he was told by his guides that it took 100,000 men, replaced every three months by fresh laborers, "ten years of oppression of the people" just to build the causeway leading to the pyramid, so that the quarried stones could be brought to the site. "The pyramid itself was twenty years in building." It was Herodotus who transmitted the information that the Pharaoh who ordered the pyramid built was Cheops (Khufu); why and what for, he does not say. Herodotus likewise attributed the Second Pyramid to Chephren (Chefra), "of the same dimensions, except that he lowered the height forty feet"; and asserted that Mycerinus (Menkara) "too left a pyramid, but much inferior in size to his father's"—implying, but not actually stating, that it was the Third Pyramid of Giza.

In the first century A.D., the Roman geographer and historian Strabo recorded not only a visit to the pyramids, but also his entry *into* the Great Pyramid through an opening in the north face, hidden by a hinged stone. Going down a long and narrow passage, he reached a pit dug in the bedrock—as other Greek and Roman tourists had done before him.

The location of this entryway was forgotten in the fol-

EXAMPLE of the CASING-STONES of a PYRAMID, SUPER-POSED
ON THE RECTANGULAR MASONRY COURSES: FROM A PHOTOGRAPH BY PS OF THE SUMMIT OF THE 2ⁿᵈ PYR.

REMNANT of the ORIGINAL CASING-STONE SURFACE of the GREAT PYRAMID.
NEAR THE MIDDLE OF ITS NORTHERN FOOT. AS DISCOVERED BY THE EXCAVATIONS OF COL HOWARD VYSE IN 1837

Fig. 130

Fig. 131

lowing centuries, and when the Moslem caliph Al Mamoon attempted to enter the pyramid in 820 A.D., he employed an army of masons, blacksmiths and engineers to pierce the stones and tunnel his way into the pyramid's core. What prompted him was both a scientific quest and a lust for treasure; for he was apprised of ancient legends that the pyramid contained a secret chamber wherein celestial maps and terrestrial spheres, as well as "weapons which do not rust" and "glass which can be bent without breaking" were hidden away in past ages.

Blasting through the mass of stones by heating and cooling them until they cracked, by ramming and chiseling, Al Mamoon's men advanced into the pyramid inch by inch. They were about to give up, when they heard the sound of a falling stone not far ahead, indicating that some cavity was located there. With renewed vigor, they blasted their way into the original Descending Passage (Fig. 132). Climbing up it, they reached the original entrance which had evaded them from the outside. Climbing down, they reached the pit described by Strabo; it was empty. A shaft from the pit led nowhere.

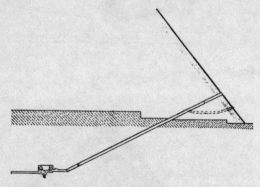

Fig. 132

As far as the searchers were concerned, the effort had been in vain. All the other pyramids, which were entered or broken into over the centuries, had the same inner structure:

a Descending Passage leading to one or more chambers. This has not been found in the Great Pyramid. There were no other secrets to be unlocked. . . .

But Fate wished otherwise. The ramming and blasting by Al Mamoon's men had loosened the stone, whose falling sound had encouraged them to tunnel on. As they were about to give up, the fallen stone was found lying in the Descending Passage. It had an odd, triangular shape. When the ceiling was examined, it was found that the stone served to hide from view a large rectangular granite slab positioned at an angle to the Descending Passage. Did it hide the way to a really secret chamber—one obviously never before visited?

Unable to move or break the granite block, Al Mamoon's men tunneled around it. It turned out that the granite slab was only one of a series of massive granite blocks, followed by limestone ones, that plugged an Ascending Passage—inclined upward at the same 26° angle that the Descending Passage was inclined downward (precisely half the angle of the pyramid's outer inclination). From the top of the Ascending Passage, a horizontal passage led to a squarish room with a gabled roof (Fig. 133) and an unusual niche in its east wall; it was bare and empty. This chamber has since been found to lie precisely in the middle of the north-south axis of the pyramid—a fact whose significance has not yet been deciphered. The chamber has come to be known as the "Queen's Chamber"; but the name is based on romantic notions and not on any shred of evidence.

At the head of the Ascending Passage, there extended for 150 feet and at the same rising angle of 26° a Grand Gallery of intricate and precise construction (Fig. 134). Its sunken floor is flanked by two ramps that run the length of the Gallery; in each ramp there are cut a series of evenly spaced rectangular slots, facing each other. The Gallery's walls rise more than 18 feet in seven corbels, each section extending three inches out above the lower one, so that the Gallery narrows as it rises. At its top, the Gallery's ceiling is the exact width as the sunken floor between the ramps.

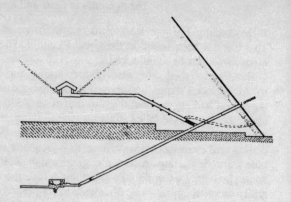

Fig. 133

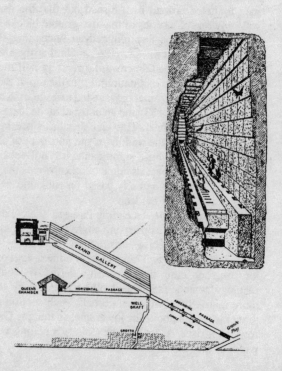

Fig. 134

At the uppermost end of the Gallery, a huge stone formed a flat platform. Flush with it a short and comparatively narrow and low corridor (only 3.5 feet high) led to an ante-chamber of extremely complex construction, having been equipped to lower with a simple maneuver (the pull of a rope?) three solid granite walls that could vertically plug the passage and block further advance.

A short corridor, of a height and width similar to the former one, then led to a high-ceilinged room constructed of red polished granite—the so-called King's Chamber (Fig. 135). It was empty except for a granite block hewed out to suggest a lidless coffer. Its precise workmanship included grooves for a lid or top section. Its measurements, as has since been determined, exhibited knowledge of profound mathematical formulas. But it was found totally empty.

Was this whole mountain of stone, then, erected to hide an empty "coffer" in an empty chamber? Blackened torch marks and the evidence of Strabo attest that the Descending Passage was visited before; if there had ever been treasure in that subterranean room, it was removed long ago. But the Ascending Passage was most definitely plugged tight when Al Ma-moon's men reached it in the ninth century A.D. The

Fig. 135

theory of the pyramids as royal tombs held that they were raised to protect the Pharaoh's mummy and the treasures buried with it from robbers and other uninvited disturbers of his eternal peace. Accordingly, the plugging of the passages is presumed to have taken place as soon as the mummy in its coffin was placed in the burial chamber. Yet here was a plugged passage—with absolutely nothing, except for an empty stone coffer, in the whole pyramid.

In time, other rulers, scientists, adventurers have entered the pyramid, tunneled and blasted through it, discovering other features of its inner structure—including two sets of shafts which some believe were air ducts (for whom?) and others assert for astronomical observations (by whom?). Although scholars persist in referring to the stone coffer as a sarcophagus (its size could well hold a human body), the fact is that there is nothing, absolutely nothing to support a claim that the Great Pyramid was a royal tomb.

Indeed, the notion that the pyramids were built as Pharaonic tombs has remained unsupported by concrete evidence.

The first pyramid, that of Zoser, contains what scholars persist in calling two burial chambers, covered by the initial mastaba. When they were first penetrated by H. M. von Minutoli in 1821, he claimed that he found inside parts of a mummy as well as a few inscriptions bearing the name of Zoser. These, it has been claimed, he sent to Europe but they were lost at sea. In 1837, Colonel Howard Vyse re-excavated the inner parts more thoroughly, and reported finding a "heap of mummies," (eighty were later counted) and to have reached a chamber "bearing the name of King Zoser," inscribed in red paint. A century later, archaeologists reported the discovery of a fragment of a skull and evidence that "a wood sarcophagus may have stood inside the red granite chamber." In 1933, J. E. Quibell and J. P. Lauer discovered beneath the pyramid additional underground galleries, in which there were two sarcophagi—empty.

It is now generally accepted that all these extra mummies and coffins represent intrusive burials, namely the entombment of the dead from a later time by intruding on the sanctity of the sealed galleries and chambers. But was Zoser himself ever entombed in the pyramid—was there ever an "original burial?"

Most archaeologists now doubt that Zoser was ever buried in the pyramid or under it. He was buried, it seems, in a magnificent tomb discovered in 1928 south of the pyramid. This "Southern Tomb," as it came to be known, was reached via a gallery whose stone ceiling imitated *palm trees*. It led to a simulated half-open door through which a great enclosure was entered. More galleries led to a subterranean room built of granite blocks; on one of its walls three false doors bore the carvings of the image, name and titles of Zoser.

Many eminent Egyptologists now believe that the pyramid was only a symbolic burial place for Zoser, and that the king was buried in the richly decorated Southern Tomb, topped by a large rectangular superstructure with a concave room which also contained the imperative chapel—just as depicted in some Egyptian drawings (Fig. 136).

The step pyramid presumed to have been begun by Zoser's

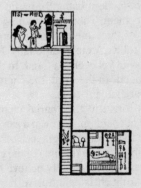

Fig. 136

successor, Sekhemkhet, also contained a "burial chamber."
It housed an alabaster "sarcophagus," which was empty.
Textbooks tell us that the archaeologist who discovered the
chamber and the stone coffer (Zakaria Goneim) concluded
that the chamber had been penetrated by grave robbers, who
stole the mummy and all other contents of the tomb; but that
is not entirely true. In fact, Mr. Goneim found the vertically
sliding door of the alabaster coffer *shut and sealed with plas-
ter,* and the remains of a dried-out wreath *still rested on top
of the coffin.* As he later recalled, "hopes were now raised to
a high pitch: but when the sarcophagus was opened, it was
found to be empty and unused." Had any king ever been buried there? While some still say yes, others are convinced that
the pyramid of Sekhemkhet (jar stoppers bearing his name
attest to the identification) was only a cenotaph (an empty,
symbolic tomb).

The third step pyramid, the one attributed to Khaba, also
contained a "burial chamber"; it was found to be completely
bare: no mummy, not even a sarcophagus. Archaeologists
have identified in the same vicinity the subterranean remains
of yet another, unfinished pyramid, believed to have been
begun by Khaba's successor. Its granite substructure contained an unusual oval "sarcophagus" sunken into the stone
floor (as an ultra-modern bathtub). Its lid was still in place,
shut tight with cement. There was nothing inside.

The remains of three other small pyramids, attributed to
Third Dynasty rulers, were additionally found. In one, the
substructure has not yet been explored. In the other, no burial
chamber was found. In the third, the chamber contained no
evidence of a burial at any time.

Nothing was found in the "burial chamber" of the collapsed pyramid of Maidum, not even a sarcophagus. Instead,
Flinders Petrie found only fragments of a wooden coffin,
which he announced as the remains of the coffin of Sneferu's
mummy. Scholars now invariably believe that it represented
the remains of a much later intrusive burial. The Maidum
pyramid is surrounded by numerous Third and Fourth Dy-

nasty mastabas, in which members of the royal family and other VIP's of that time were entombed. The pyramid's enclosure was linked with a lower structure (a so-called funerary temple) which is now submerged by the Nile's waters. It was perhaps there, surrounded and protected by the sacred river's waters, that the Pharaoh's body was laid to rest.

The next two pyramids are even more embarrassing to the pyramids-as-tombs theory. The two pyramids at Dahshur (the Bent and the Red) were both built by Sneferu. The first has *two* "burial chambers," the other *three*. All for Sneferu? If the pyramid was built by each Pharaoh to serve as his tomb, why did Sneferu build two pyramids? Needless to say, the chambers were totally empty when discovered, devoid even of sarcophagi. After some more determined excavations by the Egyptian Antiquities Service in 1947 and again in 1953 (especially in the Red Pyramid), the report admitted that "No trace of a royal tomb has been found there."

The theory of "a pyramid by each Pharaoh" now holds that the next pyramid was built by Sneferu's son, Khufu; and we have the word of Herodotus (and Roman historians who relied on his works) that it was the Great Pyramid at Giza. Its chambers, even the unviolated "King's Chamber," were empty. This should not have come as a surprise, for Herodotus (*History,* vol. II, p. 127) wrote that "the Nile water, introduced through an artificial duct, surrounds an island where the body of Cheops is said to lie." Was then the Pharaoh's real tomb somewhere lower in the valley and closer to the Nile? As of now, no one can tell.

Chefra, to whom the Second Pyramid of Giza is attributed, was not the immediate successor of Khufu. In between them a Pharaoh named Radedef reigned for eight years. For reasons which the scholars cannot explain, he selected for his pyramid a site some distance away from Giza. About half the size of the Great Pyramid, it contained the customary "burial chamber." When reached, it was found entirely empty.

The Second Pyramid of Giza has two entrances on its northern side, instead of the customary single one (see Fig.

129). The first begins—another unusual feature—outside the
pyramid and leads to an unfinished chamber. The other leads
to a chamber aligned with the pyramid's apex. When it was
entered in 1818 by Giovanni Belzoni, the granite sarcophagus was found empty and its lid lying broken on the floor. An
inscription in Arabic recorded the penetration of the chamber centuries earlier. What, if anything, the Arabs had found,
is nowhere recorded.

Giza's Third Pyramid, though much smaller than the other
two, displays many unique or unusual features. Its core was
built with the largest stone blocks of all three pyramids; its
lower sixteen courses were cased not with white limestone
but with formidable granite. It was built first as an even
smaller true pyramid (Fig. 129), then doubled in size. As a
result, it has two usable entrances; it also contains a third,
perhaps a "trial" entrance not completed by its builders.
Of its various chambers, the one deemed the main "burial
chamber" was entered in 1837 by Howard Vyse and John
Perring. They found inside the chamber a magnificently
decorated basalt sarcophagus; it was, as usual, empty. But
nearby Vyse and Perring found a fragment of a wood coffin with the royal name "Men-ka-Ra" written upon it, and
the remains of a mummy, "possibly of Menkaura"—direct
confirmation of the statement by Herodotus that the Third
Pyramid "belonged" to "Mycerinus." Modern carbon-dating
methods, however, established that the wooden coffin "certainly dates from the Saitic period"—not earlier than 660
B.C. (K. Michalowsky, *Art of Ancient Egypt);* the mummy
remains are from early Christian times. They did not belong
to any original burial.

There is some uncertainty whether Men-ka-Ra was the immediate successor of Chefra; but scholars are certain that his
successor was one named Shepsekaf. Which of the various
pyramids that were never finished (or whose construction
was so inferior that nothing remains above ground) belonged
to Shepsekaf, is still unclear. But it is certain that he was not
buried within it: he was buried under a monumental mastaba

(Fig. 137) whose burial chamber contained a black granite sarcophagus. It had been penetrated by ancient grave robbers, who emptied tomb and sarcophagus of their contents.

The Fifth Dynasty that followed began with Userkaf. He built his pyramid at Sakkara, near Zoser's pyramid complex. It was violated by both grave robbers and intrusive burials. His successor (Sahura) built a pyramid north of Sakkara (today's Abusir). Though one of the best preserved (Fig. 138), nothing was found in its rectangular "burial chamber." But the magnificence of its temples, that stretched between it and the Nile Valley, and the fact that one of the lower temple rooms was decorated with stone columns simulating palm trees, may indicate that it was somewhere near the pyramid that Sahura's real tomb was.

Neferirkara, who followed on the throne of Egypt, built his funerary complex not far from Sahara's. The chamber in his incomplete (or ruined) pyramid was empty. The monuments of his successor were not found. The next ruler built his pyramid more with dried mud bricks and wood than with stone; only meager remains of the structure were found. Neuserra, who followed, built his pyramid close by those of his predecessors. It contained two chambers—both with no trace of a burial. Neuserra, however, is better known for his funerary temple, built in the shape of a stubby, short obelisk upon a truncated pyramid (Fig. 139). The obelisk rose 118 feet; its apex was covered with gilded copper.

The pyramid of the next Pharaoh has not been found; perhaps it has crumbled to a mound, covered by the desert's shifting sands. That of his successor was identified only in 1945. Its substructure contained the usual, chamber, which was bare and empty.

The pyramid of Unash—last of the Fifth Dynasty or, as some prefer, first of the Sixth—marked a major change of custom. It was there that Gaston Maspero discovered for the first time (in 1880) the Pyramid Texts, inscribed on the walls of the pyramid's chambers and corridors. The four pyramids of the following Sixth Dynasty rulers (Teti, Pepi I,

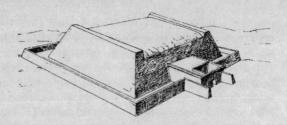

Fig. 137

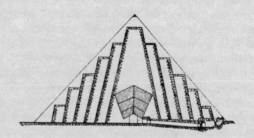

Fig. 138

Fig. 139

Mernera and Pepi II) emulated that of Unash in their funerary complexes and the inclusion of Pyramid Texts on their walls. Basalt or granite sarcophagi were found in all of their "burial" chambers; they were otherwise empty, except that in the sarcophagus in the Mernera pyramid a mummy was found. It was soon established that it was not the king's, but represented a later intrusive burial.

Where were the Sixth Dynasty kings really buried? The royal tombs of that dynasty and of earlier ones were all the way south, at Abydos. This, as the other evidence, should have completely dispelled the notion that the tombs were cenotaphs and the pyramids the real tombs; nevertheless, long-held beliefs die hard.

The facts bespeak the opposite. The Old Kingdom pyramids never held a Pharaoh's body because they were never meant to hold a king's body. In the Pharaoh's simulated Journey to the Horizon, they were built as beacons to guide his *ka* to the Stairway to Heaven—just as the pyramids originally raised by gods had served as beacons for the gods when they "sailed across the sky."

Pharaoh after Pharaoh, we suggest, attempted to emulate not the pyramid of Zoser, but the *Pyramids of the Gods:* the pyramids of Giza.

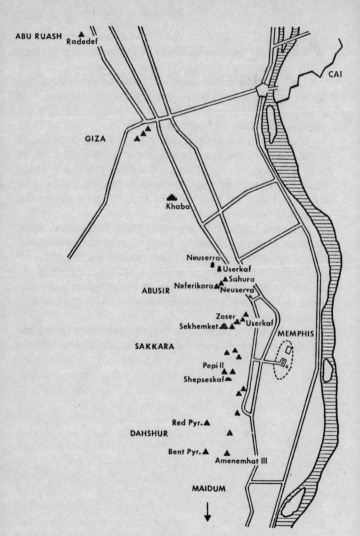

ABU RUASH ▲
Radedef

CAI

GIZA ▲▲
▲

Khaba

Neuserra ▲
▲ Userkaf
▲ Sahura
Neferikara ▲▲
▲ Neuserra
ABUSIR

Zoser ▲
Sekhemket ▲▲▲ Userkaf

MEMPHIS

SAKKARA ▲ ▲

Pepi II ▲▲
Shepseskaf ▲

▲ ▲

▲

Red Pyr. ▲
DAHSHUR ▲

Bent Pyr. ▲ ▲
Amenemhat III

MAIDUM
↓

EGYPT: THE PRINCIPAL PYRAMID SITES

Fig. 140

XIII

Forging the Pharaoh's Name

Forgery as a means to fame and fortune is not uncommon in commerce and the arts, in science and antiquities. When exposed, it may cause loss and shame. When sustained, it may change the records of history.

This, we believe, has happened to the Great Pyramid and its presumed builder, the Pharaoh named *Khufu*.

Systematic and disciplined archaeological re-examination of pyramid sites that were hurriedly excavated a century and a half ago (many times by treasure hunters), has raised numerous questions regarding some of the earlier conclusions. It has been held that the Pyramid Age began with Zoser's step pyramid, and was marked by successive progression toward a true pyramid, which finally succeeded. But why was it so important to achieve a true pyramid? If the art of pyramid building was progressively improved, why were the many pyramids which followed the Giza pyramids inferior, rather than superior to those of Giza?

Was Zoser's step pyramid the model for others, or was it itself an emulation of an earlier model? Scholars now believe that the first, smaller step pyramid (Fig. 125) that Imhotep built over the mastaba "was cased with beautiful, fine white limestone" (Ahmed Fakhry, *The Pyramids*); "before this casing was complete, however, he planned another alteration"—the superimposition of an even larger pyramid. However, as new evidence suggests, even that final step pyramid was cased, to look like a true pyramid. The casing, uncovered by archaeological missions of Harvard University led by George Reisner, was primitively made of mud bricks, which of course crumbled soon enough—leaving the impression that Zoser built a step pyramid. Moreover, these mud bricks, it was found, were whitewashed to simulate a casing of white limestone.

Whom then was Zoser trying to emulate? Where had Imhotep seen a true pyramid already up and complete, smooth sides and limestone casing and all? And another question: If, as the present theory holds, the attempts at Maidum and Sakkara to build a smooth, 52° pyramid had failed, and Sneferu had to "cheat" and build the presumed first true pyramid at an angle of only 43°—why did his son at once proceed to build a much larger pyramid at the precarious angle of 52°—and supposedly managed to achieve that with no problem at all?

If the pyramids of Giza were only "usual" pyramids in the successive chain of pyramid-per-Pharaoh—why did Khufu's son Radedef not build his pyramid next to his father's, at Giza? Remember—the other two Giza pyramids were supposedly not there yet, so Radedef had the whole site free to build as he pleased. And if his father's architects and engineers mastered the art of building the Great Pyramid, where were they to help Radedef build a similar imposing pyramid, rather than the inferior and quickly crumbling one that bears his name?

Was the reason that no other pyramid but the Great Pyramid possessed an Ascending Passage, that its unique Ascending Passage was successfully blocked and hidden until A.D. 820—so that all who emulated this pyramid knew of a Descending Passage only?

The absence of hieroglyphic inscriptions in the three pyramids of Giza is also a reason for wondering, as James Bonwick did a century ago *(Pyramid Facts and Fancies):* "Who can persuade himself that the Egyptians would have left such superb monuments without at least hieroglyphical inscriptions—they who were profuse of hieroglyphics upon all the edifices of any consideration?" The absence, one must surmise, stems from the fact that the pyramids had either been built *before* the development of hieroglyphic writing, or were not built by the Egyptians.

These are some of the points that strengthen our belief that when Zoser and his successors began the custom of pyramid building, they set out to emulate the models that had already existed: the pyramids of Giza. They were not improvements on Zoser's earlier efforts; rather, they were the prototypes which Zoser, and Pharaohs after him, attempted to emulate.

Some scholars have suggested that the small satellite pyramids at Giza were really scale models (about 1:5) that were used by the ancients exactly as today's architects use scale models for evaluation and guidance; but it is now known that they were later augmentations. However, *we think that there was indeed such a scale model; the Third Pyramid, with its obvious structural experiments. Then, we believe, the larger two were built as a pair of guiding beacons for the Anunnaki.*

But what about Menkara, Chefra and Khufu, who (we have been told by Herodotus) were the builders of these pyramids?

Well indeed—what about them? The temples and causeway attached to the Third Pyramid do bear evidence that their builder was Menkara—evidence that includes inscriptions bearing his name and several exquisite statues showing him embraced by Hathor and another goddess. But all that this attests to is that Menkara built these auxiliary structures, associating himself with the pyramid—not that he built it. The Anunnaki, it is logical to assume, needed only the pyramids and would not have built temples to worship themselves; only a Pharaoh required a funerary temple and a mortuary temple and the other structures associated with his journey to the gods.

Inside the Third Pyramid proper, not an inscription, not a statue, not a decorated wall have been found; just stark, austere precision. The only purported evidence proved to be a false pretense: the fragments of the wooden coffin inscribed with the name of Menkara proved to be from a time some 2,000 years after his reign; and the mummy "matching" the coffin was from early Christian times. There is thus not a shred of evidence to support the notion that Menkara—or any Pharaoh for that matter—had anything to do with creating and building the pyramid itself.

The Second Pyramid is likewise completely bare. Statues bearing the cartouche (oval frame within which the royal name is inscribed) of Chefra were found only in the temples adjoining the pyramid. But there is nothing at all to indicate that he had built it.

What then, about Khufu?

With one exception, *which we will expose as a probable forgery,* the only claim that he built the Great Pyramid is reported by Herodotus (and, based on his writings, by a Roman historian). Herodotus described him as a ruler who enslaved his people for thirty years to build the causeway and the pyramid. Yet by every other account, Khufu reigned for only twenty-three years. If he were such a grandiose builder, blessed with the greatest of architects and masons, where are his other monuments, where are his bigger-than-life statues?

There are none; and it would seem from the absence of such commemorative remains that Khufu was a very poor builder, not a majestic one. But he had a bright idea: our guess is that having seen the crumbled mud-brick casings of the step pyramids, the collapsed pyramid at Maidum, the hurried bending of the first pyramid of Sneferu, the improper inclination of Sneferu's second pyramid—Khufu hit upon a great idea. Out there, at Giza, there stood perfect and unspoken of pyramids. Could he not ask the gods' permission to attach to one of them the funerary temples which his Journey to the Afterlife required? There was no intrusion upon the sanctity of the pyramid itself: all the temples, including the Valley Temple in which Khufu was probably buried, were on the outside: adjoining, but not even touching, the Great Pyramid. Thus had the Great Pyramid become known as Khufu's.

Khufu's successor, Radedef, shunned his father's idea and preferred to raise his own pyramid, as Sneferu had done. But why had he gone to the north of Giza, rather than place his shrine next to his father's? The simple explanation is that the promontory of Giza was already fully occupied—by three olden pyramids plus the satellite structures erected nearby by Khufu. . . .

Witnessing Radedef's failure, the next Pharaoh—Chefra—preferred Khufu's solution. When his time came to need a pyramid, he saw no harm in appropriating for himself the ready-made second large pyramid, surrounding it with his own temples and satellites. Menkara, his successor, then attached himself to the last available pyramid, the so-called Third Pyramid.

With the ready-made pyramids thus taken, the Pharaohs who followed were forced to obtain pyramids the hard way: by trying to build them. . . . As those who had tried this before (Zoser, Sneferu, Radedef), their own efforts too ended with inferior emulations of the three olden pyramids.

At first blush, our suggestion that Khufu (as the other two) had nothing to do with building the pyramid associated with him may sound very farfetched. It is hardly so. In evidence, we call upon Khufu himself.

Whether Khufu had really built the Great Pyramid was a question that began to perplex serious Egyptologists more than a century and a quarter ago, when the *only object* mentioning Khufu and connecting him with the pyramid was discovered. Puzzlingly, it affirmed that he did not build it: *it already existed when he reigned!*

The damning evidence is a limestone stela (Fig. 141) which was discovered by Auguste Mariette in the 1850s in the ruins of the temple of Isis, near the Great Pyramid. Its inscription identifies it as a self-laudatory monument by Khufu, erected to commemorate the restoration by him of the temple of Isis and of images and emblems of the gods which Khufu found inside the crumbling temple. The opening verses unmistakably identify Khufu by his cartouche:

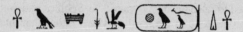

Ankh Hor Mezdau	*Suten-bat*	*Khufu tu ankh*
Live Horus Mezdau;	(To) King (of) Upper & Lower Egypt,	Khufu, is given Life!

The common opening, invoking Horus and proclaiming long life for the king, then packs explosive statements:

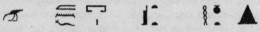

He founded   the House   of Isis, Mistress of the Pyramid,

beside the House of the Sphinx

According to the inscription on this stela (which is in the Cairo Museum), the Great Pyramid was already standing when Khufu arrived on the scene. Its mistress was the goddess Isis—it belonged to this goddess, and not to Khufu. Furthermore, the Sphinx too—which has been attributed to Chefra, who presumably built it together with the Second Pyramid—was also already crouching at its present location. The continuation of the inscription pinpoints the position of the Sphinx accurately, and records the fact that part of it was damaged by lightning—a damage perceivable to this very day.

Khufu continues to state in his inscription that he built a pyramid for the Princess Henutsen "beside the temple of the goddess." Archaeologists have found independent evidence that the southernmost of the three small pyramids flanking

Fig. 141

the Great Pyramid—the small pyramid nearest the temple of Isis—was in fact dedicated to Henutsen, a wife of Khufu. Everything in the inscription thus matches the known facts; but the only pyramid-building claim made by Khufu is that he built the small pyramid for the princess. The Great Pyramid, he states, was already there, as was the Sphinx (and, by inference, the other two pyramids as well).

Such support for our theories is even further strengthened, as we read in another portion of the inscription that the Great Pyramid was also called "The Western Mountain of Hathor":

> Live Horus Mezdau;
> To King of Upper and Lower Egypt, Khufu,
> Life is given.
> For his mother Isis, the Divine Mother,
> Mistress of "The Western Mountain of Hathor,"
>   he made (this) writing on a stela.
> He gave (her) a new sacred offering.
> He built (her) a House (temple) of stone,
>   renewed the gods that were found in her temple.

Hathor, we will recall, was the mistress of the Sinai peninsula. If the highest peak of the peninsula was her Eastern Mountain, the Great Pyramid was her Western Mountain—the two acting as the anchors for the Landing Corridor.

This "Inventory Stela," as it came to be called, bears all the marks of authenticity. Yet scholars at the time of its discovery (and many ever since) have been unable to reconcile themselves to its unavoidable conclusions. Unwilling to upset the whole structure of Pyramidology, they proclaimed the Inventory Stela a *forgery*—an inscription made "long after the death of Khufu" (to quote Selim Hassan, *Excavations at Giza),* but invoking his name "to support some fictitious claim of the local priests."

James H. Breasted, whose *Ancient Records of Egypt* is the standard work on ancient Egyptian inscriptions, wrote in 1906 that "the references to the Sphinx, and the so-called temple beside it in the time of Khufu, have made this monument from the first an object of great interest. These references would be of the highest importance if the monument were contemporaneous with Khufu; but the orthographic evidences of its late date are entirely conclusive." He disagreed with Gaston Maspero, a leading Egyptologist of the time, who had earlier suggested that the stela, if indeed of late orthography, was a copy of an earlier and authentic original. In spite of the doubts, Breasted included the inscription among the records of the Fourth Dynasty. And Maspero, when he wrote his comprehensive *The Dawn of Civilization* in 1920, accepted the contents of the Inventory Stela as factual data concerning the life and activities of Khufu.

Why then the reluctance to call the artifact authentic?

The Inventory Stela was condemned as a forgery because only a decade or so earlier the identification of Khufu as the builder of the Great Pyramid appeared to have been undisputably established. The seemingly conclusive evidence was markings in red paint, discovered in sealed chambers above the King's Chamber, which could be interpreted as masons' markings made in the eighteenth year of the reign of Khufu (Fig. 142). Since the chambers were not entered until discovered in 1837, the markings must have been authentic; and if the Inventory Stela offered contradictory information, the Stela must have been a forgery.

But as we probe the circumstances of the red-paint markings, and ascertain who the discoverers were—an inquiry somehow never undertaken before—the conclusion that emerges is this: if a forgery had taken place, it occurred not in ancient times but in the year A.D. 1837; and the forgers were not "some local priests," but two (or three) unscrupulous Englishmen. . . .

The story begins with the arrival in Egypt on December

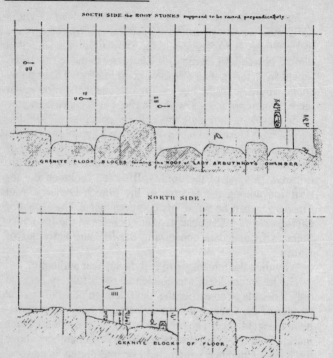

Fig. 142

29, 1835 of Colonel Richard Howard Vyse, a "black sheep" of an aristocratic English family. At that time, other officers of Her Majesty's Army had become prominent in the ranks of "antiquarians" (as archaeologists were then called), reading papers before distinguished societies and receiving due public accolade. Whether or not Vyse had gone to Egypt with such notions in mind, the fact is that visiting the pyramids of Giza, he was at once caught by the fever of daily discoveries by scholars and laymen alike. He was especially thrilled by the tales and theories of one Giovanni Battista Caviglia, who had been searching for a hidden chamber inside the Great Pyramid.

Within days, Vyse offered to provide the funds for Caviglia's search, if he were accepted as a co-discoverer. Caviglia rejected the offer outright; and the offended Vyse sailed off to Beirut at the end of February 1836, to visit Syria and Asia Minor.

But the long trip did not cure the craving that was aroused within him. Instead of returning to England, he showed up back in Egypt in October 1836. On the earlier visit, he had befriended a crafty go-between by the name of J. R. Hill, then a copper mill superintendent. Now he was introduced to a "Mr. Sloane," who whispered that there were ways to get a *Firman*—a concession decree—from the Egyptian government to sole excavation rights at Giza. Thus guided, Vyse went to the British Consul, Col. Campbell, for the necessary documentation. To his great shock, the Firman named Campbell and Sloane as co-permitees, and designated Caviglia as the works' supervisor. On November 2, 1836, the disappointed Vyse paid over to Caviglia "my first subscription of 200 dollars" and left in disgust on a sightseeing trip to Upper Egypt.

As chronicled by Vyse in his *Operations Carried on at the Pyramids of Gizeh in 1837,* he returned to Giza on January 24, 1837, "extremely anxious to see what progress had been made." But instead of searching for the hidden chamber, Caviglia and his workmen were busy digging up mummies from tombs around the pyramids. Vyse's fury subsided only when Caviglia asserted that he had something important to show him: writing by the pyramids' builders!

The excavations at the tombs showed that the ancient masons sometimes marked the pre-cut stones with red paint. Such markings, Caviglia said, he found at the base of the Second Pyramid. But when examined with Vyse, the "red paint" turned out to be natural discolorations in the stone.

What about the Great Pyramid? Caviglia, working there to discover where the "air channels" were leading from the

"King's Chamber," was more than ever convinced that there were secret chambers higher up. One such compartment, reachable via a crawlway, was discovered by Nathaniel Davison in 1765 (Fig. 143). Vyse demanded that work be concentrated there; he was dismayed to find out that Caviglia and Campbell were more interested in finding mummies, which every museum then desired. Caviglia had even gone so far as to name a large tomb he had found "Campbell's Tomb."

Determined to run his own show, Vyse moved from Cairo to the site of the pyramids. "I naturally wished to make some discoveries before I returned to England," he admitted in his journal on January 27, 1837. At great expense to his family, he was now gone for well over a year.

In the following weeks, the rift with Caviglia widened as Vyse hurled at him various accusations. On February 11, the two had a violent argument. On the twelfth, Caviglia made major discoveries in Campbell's Tomb: a sarcophagus inscribed with hieroglyphs and masons' red-paint markings on the stone walls of the tomb. On the thirteenth, Vyse summarily discharged Caviglia and ordered him away from the site. Caviglia returned only once, on the fifteenth, to pick up his

Fig. 143

belongings; for years thereafter, he made "dishonorable accusations" against Vyse, whose nature Vyse's chronicles do not care to detail.

Was the row a genuine disagreement, or did Vyse artificially bring matters to a head in order to get Caviglia off the site?

As it turned out, Vyse secretly entered the Great Pyramid on the night of February 12, accompanied by one John Perring—an engineer with the Egyptian Public Works Department and a dabbler in Egyptology—whom Vyse met through the resourceful Mr. Hill. The two examined an intriguing crevice that had developed in a granite block above Davison's Chamber; when a reed was pushed in, it went through unbent; there was obviously some space beyond.

What schemes did the two concoct during that secret night visit? We can only guess from future events. The facts are that Vyse dismissed Caviglia the next morning and put Perring on his payroll. In his journal, Vyse confided: "I am determined to carry on the excavations above the roof of (Davison's) Chamber, where I expect to find a sepulchral apartment." As Vyse threw more men and money behind this search, royalty and other dignitaries came to inspect the finds at Campbell's Tomb; there was little new that Vyse could show them inside the pyramid. In frustration, Vyse ordered his men to bore into the shoulder of the Sphinx, hoping to find its masons' markings. Unsuccessful, he refocused his attention on the Hidden Chamber.

By mid-March, Vyse faced a new problem: other projects were luring away his workmen. He doubled their pay, if only they would work day and night: time, he realized, was running out. In desperation, Vyse threw caution to the winds, and ordered the use of explosives to blast his way through the stones that blocked his progress.

By March 27, the workmen managed to cut a small hole through the granite slabs. Illogically, Vyse thereupon discharged the foreman, one named Paulo. On the following day, Vyse wrote, "I inserted a candle at the end of a rod through a small hole that had been made in the chamber above Davison's, and I had the mortification of finding that it was a chamber of construction like that below it." He had found the Hidden Chamber! (Fig. 144.)

Using gunpowder to enlarge the hole, Vyse entered the newly discovered chamber on March 30—accompanied by Mr. Hill. They examined it thoroughly. It was hermetically sealed, with no opening whatsoever. Its floor consisted of the rough side of the large granite slabs that formed the ceiling of Davison's Chamber below. "A black sediment was equally distributed all over the floor, showing each footstep." (The nature of this black powder, which was "accumulated to some depth," has never been ascertained.) "The ceiling was beautifully polished and had the finest joints." The chamber, it was clear, had never been entered before; yet it contained neither sarcophagus nor treasure. It was bare—completely empty.

Vyse ordered the hole enlarged, and sent a message to the British Consul announcing that he had named the new

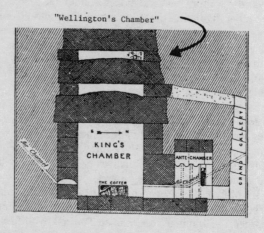

Fig. 144

compartment "Wellington's Chamber." In the evening, "Mr. Perring and Mr. Mash having arrived, we went into Wellington's Chamber and took various measurements, *and in doing so we found the quarry marks.*" What a sudden stroke of luck!

They were similar to the red-painted quarry marks found in tombs outside the pyramid. Somehow, Vyse and Hill missed them entirely when they thoroughly inspected the chamber by themselves. But joined by Mr. Perring and by Mr. Mash—a civil engineer who was present at Perring's invitation—there were four witnesses to the unique discovery.

The fact that Wellington's Chamber was almost identical to Davison's led Vyse to suspect that there was yet another chamber above it. For no given reason Vyse dismissed on April 4 the remaining foreman, one named Giachino. On April 14, the British Consul and the Austrian Consul General visited the site. They requested that copies be made of the masons' markings. Vyse put Perring and Mash to work—but instructed them to copy first the earlier-discovered markings in Campbell's tomb; the unique ones inside the Great Pyramid could somehow wait.

With liberal use of gunpowder, the compartment above Wellington's (Vyse named it after Lord Nelson) was broken into on April 25. It was as empty as the others, its floor also covered with the mysterious black dust. Vyse reported that he found "several quarry marks inscribed in red upon the blocks, particularly on the west side." All along, Mr. Hill was going in and out of the newly found chambers, ostensibly to inscribe in them (how?) the names of Wellington and Nelson. On the twenty-seventh Mr. Hill—not Perring or Mash—copied the quarry marks. Vyse reproduced the ones from Nelson's Chamber (though not the ones from Wellington's) in his book (Fig. 145a).

On May 7, the way was blasted through into one more chamber above Nelson's, which Vyse named temporarily after Lady Arbuthnot. The journal entry makes no mention of any quarry marks, although they were later on found there in profusion. What was striking about the new markings was that they included cartouches—which could only mean royal names (Fig. 145b)—in profusion. Has Vyse come upon the actual written name of the Pharaoh who had built the pyramid?

On May 18, a Dr. Walni "applied for copies of the characters found in the Great Pyramid, in order to send them to Mr. Rosellini," an Egyptologist who had specialized in the decipherment of royal names. Vyse turned the request down outrightly.

The next day, in the company of Lord Arbuthnot, a Mr. Brethel and a Mr. Raven, Vyse entered Lady Arbuthnot's Chamber and the four "compared Mr. Hill's drawings with the quarry marks in the Great Pyramid; and we afterward signed an attestation to their accuracy." Soon thereafter, the final vaulted chamber was broken into, and more markings—including a royal cartouche—were found. Vyse then proceeded to Cairo and submitted the authenticated copies of the writings on the stones to the British Embassy, for official forwarding to London.

His work was done: he found hitherto unknown chambers, and he proved the identity of the builder of the Great Pyramid; for within the cartouches was written the royal name *Kh-u-f-u* (⊙ 𓏲 ⸺ 𓏲)

To this discovery, every textbook has been attesting to this very day.

The impact of Vyse's discoveries was great, and their acceptance assured, after he managed to quickly obtain a confirmation from the experts of the British Museum in London.

When the facsimiles made by Mr. Hill reached the Museum, and when exactly their analysis reached Vyse, is not clear; but he made the Museum's opinion (by the hand of its hieroglyphics expert Samuel Birch) part of his chronicle of May 27, 1837. On the face of it, the long analysis confirmed Vyse's expectations: the names in the cartouches could be

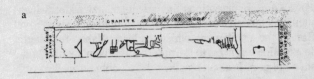

a

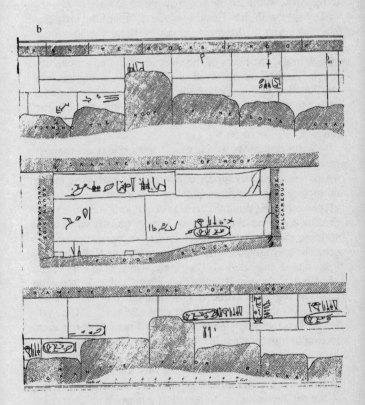

Fig. 145

read as *Khufu* or variations thereof: just as Herodotus had written, Cheops was the builder of the Great Pyramid.

But in the excitement which understandably followed, little attention was paid to the many if's and but's in the Museum's opinion. It also contained the clue that tipped us off to the forgery: the forger's clumsy mistake.

To begin with, Mr. Birch was uneasy about the orthography and script of the many markings. "The symbols or hieroglpyhs traced in red by the sculptor, or mason, upon the stones in the chambers of the Great Pyramid are apparently quarry marks," he observed in his opening paragraph; the qualification at once followed: "Although not very legible, owing to their having been written in semi-hieratic or linear-hieroglyphic characters, they possess points of considerable interest. . . ."

What puzzled Mr. Birch was that markings presumably from the beginning of the Fourth Dynasty were made in a script that started to appear only centuries later. Originating as pictographs—"written pictures"—the writing of hieroglyphic symbols required great skill and long training; so, in time, in commercial transactions, a more quickly written and simpler, more linear script referred to as hieratic came into use. The hieroglyphic symbols discovered by Vyse thus belonged to another period. They were also very indistinct and Mr. Birch had great difficulty in reading them: "The meaning of the hieroglyphics following the prenomen in the same linear hand as the cartouche, is not very obvious. . . . The symbols following the name are very indistinct." Many of them looked to him "written in characters very nearly hieratic"— from an even much later period than the semi-hieratic characters. Some of the symbols were very unusual, never seen in any other inscription in Egypt: "The cartouche of Suphis" (Cheops), he wrote, "is followed by a hieroglyphic to which it would be difficult to find a parallel." Other symbols were "equally difficult of solution."

Mr. Birch was also puzzled by "a curious sequence of symbols" in the upper-most, vaulted chamber (named by

Vyse "Campbell's Chamber"). There, the hieroglyphic symbol for "good, gracious" was used as a numeral—a usage never discovered before or since. Those unusually written numerals were assumed to mean "eighteenth year" (of Khufu's reign).

No less puzzling to him were the symbols which followed the royal cartouche and which were "in the same linear hand as the cartouche." He assumed that they spelled out a royal title, such as "Mighty in Upper and Lower Egypt." The only similarity that he could find to this row of symbols was that of "a title that appears on the coffin of the queen of Amasis" of the Saitic period. He saw no need to stress that the Pharaoh Amasis had reigned in the sixth century B.C.—more than 2,000 years after Khufu!

Whoever daubed the red-paint markings reported by Vyse had thus employed a writing method (linear), scripts (semi-hieratic and hieratic) and titles from various periods—but none from the time of Khufu, and all from later periods. Their writer was also not too literate: many of his hieroglyphs were either unclear, incomplete, out of place, erroneously employed or completely unknown.

(Analyzing these inscriptions a year later, the leading German Egyptologist of the time, Carl Richard Lepsius, was likewise puzzled by the fact that the inscriptions "were traced with a brush in red paint in a cursive manner, so much so that they resemble hieratic signs." Some of the hieroglyphs following the cartouches, he declared, were totally unknown, and "I am unable to explain them.")

Turning to the main issue on which he was requested to give an opinion—the identity of the Pharaoh named in the inscriptions—Birch threw a bombshell: there were *two,* and not just one, royal names within the pyramid!

Was it possible that two kings had built the same pyramid? And if so, who were they?

The two royal names appearing in the inscriptions, Samuel Birch reported, were not unknown: "they had already been found in the tombs of functionaries employed by monarchs

of that dynasty," namely the Fourth Dynasty to whose Pharaohs the pyramids of Giza were attributed. One cartouche
(Fig. 146a) was then read *Saufou* or *Shoufou;* the other
(146b) included the ram symbol of the god Khnum and was
then read *Senekhuf* or *Seneshoufou.*

Attempting to analyze the meaning of the name with the
ram symbol, Birch noted that "a cartouche, similar to that
which first occurs in Wellington's Chamber, had been published by Mr. Wilkinson, mater. Hieroglyph, Plate of Un-
placed Kings E; and also by Mr. Rosellini, tom. i. tav.1,3,
who reads the phonetic elements of which it is composed
'Seneshufo,' which name is supposed by Mr. Wilkinson to
mean 'the Brother of Suphis.'"

That one Pharaoh might have completed a pyramid begun
by his predecessor has been a theory accepted by Egyptologists (as in the case of the pyramid at Maidum). Could not
this account for two royal names within the same pyramid?
Perhaps—but certainly not in our case.

The impossibility in the case of the Great Pyramid stems
from the location of the various cartouches (Fig. 147). The
cartouche that is presumed to have belonged in the pyramid, that of Cheops/Khufu, was found only in the *uppermost,* vaulted chamber, the one named Campbell's Chamber.
The several cartouches which spelled out the second name
(nowadays read *Khnem-khuf*) appeared in Wellington's
Chamber and in Lady Arbuthnot's Chamber (no cartouches
were inscribed in Nelson's Chamber). In other words, the
lower chambers bore the name of a Pharaoh who lived and
reigned *after* Cheops. As there was no way to build the pyramid except from its base upward, the location of the cartouches meant that Cheops, who reigned before Chephren,
completed a pyramid begun by a Pharaoh who succeeded
him. That, of course, was not possible.

Conceding that the two names could have stood for what
the ancient King Lists had called Suphis I (Cheops) and
Suphis II (Chephren), Birch tried to resolve the problem
by wondering whether both names, somehow, belonged to

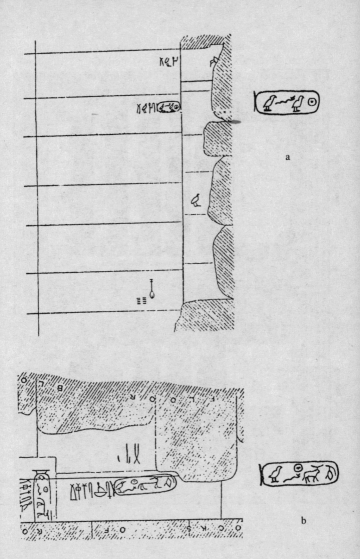

Fig. 146

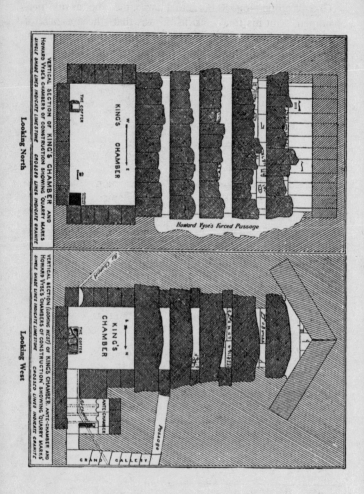

VERTICAL SECTION OF KING'S CHAMBER AND HOWARD VYSE'S CHAMBERS OF CONSTRUCTION SHOWING QUARRY MARKS
SINGLE SHADE LINES INDICATE LIMESTONE   CROSSED LINES INDICATE GRANITE

Looking North

THE COFFER

KING'S CHAMBER

W ——— E

Howard Vyse's Forced Passage

VERTICAL SECTION (LOOKING WEST) OF KING'S CHAMBER, ANTE-CHAMBER AND HOWARD VYSE'S CHAMBERS OF CONSTRUCTION SHOWING QUARRY MARKS
SINGLE SHADE LINES INDICATE LIMESTONE   CROSSED LINES INDICATE GRANITE

Looking West

THE COFFER

KING'S CHAMBER

S ——— N

ANTE-CHAMBER

Passage

GRAND GALLERY

Fig. 147

Cheops alone—one as his actual name, the other as his "pre-nomen." But his final conclusion was that "the presence of this (second) name, as a quarry-mark, in the Great Pyramid, is an additional embarrassment" on top of the other embarrassing features of the inscriptions.

The "Problem of the Second Name" was still unresolved when England's most noted Egyptologist, Flinders Petrie, spent months measuring the pyramids a half century later. "The most destructive theory about this king (Khnem-khuf) is that he is identical with Khufu," Petrie wrote in *The Pyramids and Temples of Gizeh,* giving the many reasons voiced by then by other Egyptologists against such a theory. For any number of reasons, Petrie showed, the two names belonged to two separate kings. Why then did both names appear within the Great Pyramid in the locations in which they did? Petrie believed that the only plausible explanation would be that Cheops and Chephren were co-regents, reigning together.

Since no evidence to support Petrie's theory has been found, Gaston Maspero wrote almost a century after the discovery by Vyse that "the existence of the two cartouches Khufui and Khnem-Khufui on the same monuments has caused much embarrassment to Egyptologists" *(The Dawn of Civilization).* The problem, in spite of all suggested solutions, is still an embarrassing one.

But a solution, we believe, can be offered—if we stop attributing the inscriptions to ancient masons, and begin to look at the facts.

The pyramids of Giza are unique, among other things, for the complete absence of any decoration or inscription within them—with the outstanding exception of the inscriptions found by Vyse. Why the exception? If the masons felt no qualms about daubing in red paint inscriptions upon the blocks of stones hidden away in the compartments above the "King's Chamber," why were there absolutely no such inscriptions found in the first compartment, the one discovered by Davison in 1765—but only in the compartments found by Vyse?

In addition to the inscriptions reported by Vyse, there have been found in the various compartments true masons' markings—positioning lines and arrows. They are all drawn as one would expect, with the right side up; for when they were drawn, the compartment in which the masons worked was not yet roofed: they could stand up, move about and draw the markings without encumberment. But all the inscriptions—drawn over and around the masons' markings (Fig. 145)—are either *upside down* or vertical, as though whoever drew them had to bend or crouch within the low compartments (their height varied from one foot four inches to four feet five inches in Lady Arbuthnot's Chamber, from two feet two inches to three feet eight inches in Wellington's Chamber).

The cartouches and royal titles daubed upon the walls of the compartments were imprecise, crude and extra large. Most cartouches were two and a half to three feet long and about a foot wide, sometimes occupying the better part of the face of the stone block on which they were painted—as though the inscriber had needed all the space he could get. They are in sharp contrast to the precision and delicacy and perfect sense of proportion of ancient Egyptian hieroglyphics, evident in the true masons' markings found in those same compartments.

With the exception of a few markings on a corner of the eastern wall in Wellington's Chamber, no inscriptions were found on the eastern walls of any other chamber; nor were there any other symbols (other than the original masons' markings) found on any of these other eastern walls, except for a few meaningless lines and a partial outline of a bird on the vaulted eastern end of Campbell's Chamber.

This is odd, especially if one realizes that it was from the eastern side that Vyse had tunneled to and broken into these compartments. Did the ancient masons anticipate that Vyse would break in through the eastern walls, and obliged by not putting inscriptions on them? Or does the absence of such inscriptions suggest that whoever daubed them preferred to write on the intact walls to the north, south and west, rather than on the damaged east walls?

In other words: cannot all the puzzles be solved, if we assume that the inscriptions were not made in antiquity, when the pyramid was being built, but only *after* Vyse had blasted his way into the compartments?

The atmosphere that surrounded Vyse's operations in those hectic days is well described by the Colonel himself. Major discoveries were being made all around the pyramids, but not within them. Campbell's Tomb, discovered by the detested Caviglia, was yielding not only artifacts but also masons' markings and hieroglyphics in red paint. Vyse was becoming desperate to achieve his own discovery. Finally he broke through to hitherto unknown chambers; but they only duplicated one after the other a previously discovered chamber (Davison's) and were bare and empty. What could he show for all the effort and expenditure? For what would he be honored, by what would he be remembered?

We know from Vyse's chronicles that, by day, he had sent in Mr. Hill to inscribe the chambers with the names of the Duke of Wellington and Admiral Nelson, heroes of the victories over Napoleon. By night, we suspect, Mr. Hill also entered the chambers—to "christen" the pyramid with the cartouches of its presumed ancient builder.

"The two royal names," Birch pointed out in his Opinion, "had already been found in the tombs of functionaries employed by the monarchs of that dynasty under which these Pyramids were erected." The Pharaoh's artisans surely knew the correct name of their king. But in the 1830s Egyptology was still in its infancy; and no one could yet tell for sure which was the correct hieroglyphic design of the king whom Herodotus called "Cheops."

And so it was, we suspect, that Mr. Hill—probably alone, certainly at night when all others were gone—had entered the newly discovered chambers. Using the imperative red paint, by torchlight, crouching and bending in the low compartments, he strained to copy hieroglyphic symbols from some source; and he drew on the walls that were intact what seemed to him appropriate markings. He ended up inscribing, in Wellington's Chamber as in Lady Arbuthnot's, the wrong name.

With inscriptions of royal names of the Fourth Dynasty popping up in the tombs surrounding the pyramids of Giza, which were the right cartouches to be inscribed by Hill? Unschooled in hieroglyphic writing, he must have taken with him into the pyramid some source book from which to copy the intricate symbols. The one and only book repeatedly mentioned in Vyse's chronicles is (Sir) John Gardner Wilkinson's *Materia Hieroglyphica.* As its title page declared, it aimed to update the reader on "the Egyptian Pantheon and the Succession of the Pharaohs from the earliest times to the conquest of Alexander." Published in 1828—nine years before Vyse's assault on the pyramids—it was a standard book for English Egyptologists.

Birch had stated in his report, "a cartouche, similar to that which first occurs in Wellington's Chamber, had been published by Mr. Wilkinson *Mater. Hieroglyph.*" We thus have a clear indication of the probable source of the cartouche inscribed by Hill in the very first chamber (Wellington's) found by Vyse (Fig. 146b).

Having looked up Wilkinson's *Materia Hieroglyphica,* we can sympathize with Vyse and Hill: its text and presentation are disorganized, and its plates reproducing cartouches are small, ill-copied and badly printed. Wilkinson appears to have been uncertain not only regarding the reading of royal names, but also regarding the correct manner by which hieroglyphs carved or sculpted on stone should be transcribed in linear writing. The problem was most acute concerning the disk sign, which on such monuments appeared as either a solid disk ● or as a void sphere ○ , and in linear (or brushed-on) writing as a circle with a dot in its center ⊙ . In his works, he transcribed the royal cartouches in question in some instances as a solid disk, and in others as a circle with a dot in its center.

Hill had followed Wilkinson's guidance. But all of these cartouches were of the *Khnum* variety. Timewise, it means that by May 7 only the "ram" cartouches were inscribed. Then on May 27, when Campbell's Chamber was broken into, the vital and conclusive cartouche spelling Kh-u-f-u was found. How did the miracle happen?

A clue is hidden in a suspicious segment in Vyse's chronicles, in an entry devoted to the fact that the casing stones "did not show the slightest trace of inscription or of sculpture, nor, indeed, was any to be found upon any stone belonging to the pyramid, or near it (with the exception of the quarry-marks already described)." Vyse noted that there was one other exception: "part of a cartouche of Suphis, engraved on a brown stone, six inches long by four broad. This fragment was dug out of the mound at the northern side on June 2." Vyse reproduced a sketch of the fragment (Fig. 148a).

How did Vyse know—even before the communication from the British Museum—that this was "part of a cartouche of *Suphis*?" Vyse would like us to believe it was because a week earlier (on May 27) he had found the complete cartouche (Fig. 148b) in Campbell's Chamber.

But here is the suspicious aspect. Vyse claims in the above-quoted entry that the stone with the partial Khufu cartouche was found on *June 2*. Yet his entry is dated *May 9*! Vyse's manipulation of dates would have us believe that the partial cartouche found outside the pyramid corroborated the earlier find of the complete cartouche inside the pyramid.

a   b

Fig. 148

But the dates suggest that it was the other way around: Vyse had already realized on May 9—a full eighteen days *before* the discovery of Campbell's Chamber—what the crucial cartouche had to look like. Somehow, on May 9, Vyse and Hill had realized that they had missed out on the correct name of Cheops.

This realization could explain the frantic, daily commuting by Vyse and Hill to Cairo right after the discovery of Lady Arbuthnot's Chamber. Why they had left when so badly needed at the pyramids, the Chronicles do not state. We believe that the "bombshell" that hit them was yet another, new work by Wilkinson, the three-volume *Manners and Customs of the Ancient Egyptians*. Published in London earlier that year (1837), it must have reached Cairo right during those dramatic and tense days. And, neatly and clearly printed for a change, it reproduced in a chapter on early sculptures both the ram cartouche which Vyse-Hill had already copied—and a new cartouche, one which Wilkinson read "Shufu or Suphis" (Fig. 149).

Wilkinson's new presentation must have shocked Vyse and Hill, because he appeared to have changed his mind regarding the ram cartouche (No. 2 in his Plate). He now read it "Numba-khufu or Chembes" rather than "Sen-Suphis." These names, he wrote, were found inscribed in tombs in

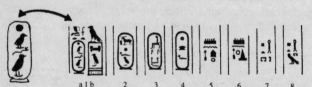

a 1 b   2   3   4   5   6   7   8

1. a, b, the name Shufu, or Suphis. 2. Numba-khufu, or Chembes. 3 Asseskas, or Shpeskaf.
4. Shafra, Khafru, or Kephren.   5, 6. The name of Memphis.
7, 8. (Memphis, or Ptah-ei, the abode of Ptah.)

From the Tombs near the Pyramids.

Fig. 149

the vicinity of the Great Pyramid; and it was in cartouche la that "we perceive Suphis, or, as the hieroglyphics wrote it, Shufu or Khufu, a name easily converted into Suphis or Cheops." So *that* was the correct name that had to be inscribed!

For whom, then, did the ram cartouche (his fig. 2) stand? Explaining the difficulties of identification, Wilkinson admitted that he could not decide "whether the first two names here introduced are both of Suphis, or if the second one is of the founder of the other pyramid."

With this unsettling news, what were Vyse and Hill to do? Wilkinson's narrative gave them a lead, which they hurried to follow. The two names, he wrote on, "occur again at Mount Sinai."

Somewhat inaccurately—a fault common in his work— Wilkinson was referring to hieroglyphic inscriptions found not actually at Mount Sinai, but in the Sinai's area of the turquoise mines. The inscriptions became known in those years due to the magnificently illustrated *Voyage de l'Arabie Pétrée* in which Léon de Laborde et Linat described the Sinai peninsula. Published in 1832, its drawings included reproductions of monuments and inscriptions in the wadi leading to the mining area, Wadi Maghara. There, Pharaoh after Pharaoh carved on the rocks mementoes of their achievements in holding the mines against marauding Asiatics. One such depiction (Fig. 150) included the two cartouches of which Wilkinson wrote.

Vyse and Hill should have had little difficulty in locating a copy of de Laborde's *Voyage* in French-speaking Cairo. The particular drawing seemed to answer Wilkinson's doubt: the same Pharaoh appeared to have two names, one with the ram symbol and the other that spelt out Kh-u-f-u. Thus, by May 9, Vyse-Hill-Perring had learned that one more cartouche was needed, and what it had to look like.

When Campbell's Chamber was broken into on May 27, the three must have asked themselves: what are we waiting for? And so it was that the final conclusive cartouche

Fig. 150

appeared on the uppermost wall (Fig. 146a). Fame, if not fortune, was assured for Vyse; Mr. Hill, on his part, did not come out of the adventure empty-handed.

How sure can we be of our accusations, a century and a half after the event?

Sure enough. For, as most forgers, Mr. Hill made, on top of all the other embarrassments, one grave mistake: a mistake that no ancient scribe could have possibly committed.

As it turned out, both source books by which Vyse-Hill were guided (Wilkinson's *Materia Hieroglyphica* and then de Laborde's *Voyage*) contained spelling errors; the unsuspecting team embodied the errors in the pyramid's inscriptions.

Samuel Birch himself pointed out in his report that the hieroglyph for *Kh* (the first consonant in the name Kh-u-f-u), which is ◉ (representing pictorially a sieve), "appears in Mr. Wilkinson's work without distinction from the solar disk." The *Kh* hieroglyph had to be employed in all the cartouches (spelling Khnem-*Kh*-u-f) which were inscribed in the two lower chambers. *But the correct sieve symbol was not employed even once.* Instead, the consonant *Kh* was represented by the symbol for the Solar Disk: whoever inscribed these cartouches made the same error as Wilkinson had made. . . .

When Vyse and Hill got hold of de Laborde's book, its sketch only deepened the error. The rock carvings depicted by him included the cartouche Kh-u-f-u on the right, and Khnum-kh-u-f on the left. In both instances, de Laborde—who admitted to ignorance of hieroglyphics and who made no attempt to read the symbols—rendered the *Kh* sign as a void circle ○ (see Fig. 150). (The *Kh* symbol was correctly spelled ◉ in the rock carvings, as has been verified by all scholarly authorities—viz. Lepsius in *Denkmäler,* Kurt Sethe in *Urkunden des Alten Reich,* and *The Inscriptions of Sinai* by A. H. Gardiner and T. E. Peet. De Laborde made another fateful error: He depicted as one Pharaoh's inscription, with two royal names, what were in effect *two* adjoining inscriptions, in different script styles, by two different Pharaohs—as is clearly seen in Fig. 151).

Fig. 151

His depiction thus served to enhance Vyse's and Hill's notion that the crucial cartouche of Kh-u-f-u should be inscribed in the uppermost chamber with the symbol for the Solar Disk (146a). *But in doing so, the inscriber had employed the hieroglyphic symbol and phonetic sound for RA, the supreme god of Egypt!* He had unwittingly spelled out not *Khnem-Khuf,* but *Khnem-Rauf;* not *Khufu,* but *Raufu.* He had used the name of the great god incorrectly and in vain; it was blasphemy in ancient Egypt.

It was also an error inconceivable for an Egyptian scribe of the times of the Pharaohs. As monument after monument and inscription after inscription make clear, the symbol for *Ra* ☉ and the symbol for *Kh* ◉ were always correctly employed—not only in different inscriptions, but also in the same inscription by the same scribe.

And, therefore, the substitution of *Ra* for *Kh* was an error that could not have been committed in the time of Khufu, nor of any other ancient Pharaoh. Only a stranger to hieroglyphics, a stranger to Khufu, and a stranger to the overpowering worship of Ra, could have committed such a grave error.

Added to all the other puzzling or inexplicable aspects of the discovery reported by Vyse, this final mistake establishes conclusively, we believe, that Vyse and his aides, not the original builders of the Great Pyramid, caused the red-painted markings to be inscribed.

But, one may ask, was there no risk that outside visitors— such as the British and Austrian consuls, or Lord and Lady Arbuthnot—would notice that the inscriptions were so much fresher-looking than the masons' true markings? The question was answered at the time by one of the men involved, Mr. Perring, in his own volume on the subject *(The Pyramids of Gizeh).* The paint used for the ancient inscriptions, he wrote, was a "composition of red ochre called by the Arabs *moghrah* (which) is still in use." Not only was the same red ochre paint available, Perring stated, but "such is the state of preservation of the marks in the quarries, that *it is difficult to distinguish the mark of yesterday from one of three thousand years."*

The forgers, in other words, were sure of their ink.

Were Vyse and Hill—possibly with the tacit connivance of Perring—morally capable of perpetrating such a forgery?

The circumstances of Vyse launching into this adventure of discovery, his treatment of Caviglia, the chronology of events, his determination to obtain a major find as time and money were running out—bespeak a character capable of such a deed. As to Mr. Hill—whom Vyse endlessly thanks in his foreword—the fact is that having been a copper mill employee when he first met Vyse, he ended up owning the Cairo Hotel when Vyse left Egypt. And as to Mr. Perring, a civil engineer turned Egyptologist—well, let subsequent events speak for themselves. For, encouraged by the success of one forgery, the Vyse team attempted one and probably two more. . . .

All along, as the discoveries were being made in the Great Pyramid, Vyse half-heartedly continued Caviglia's work in and around the other two pyramids. Encouraged by his newly won fame by the Great Pyramid discoveries, Vyse decided to postpone his return to England and instead engage in concerted efforts to uncover the secrets of the other two pyramids.

With the exception of red-painted markings on stones, which experts from Cairo determined were from tombs or structures outside the pyramids and not from within the pyramids, nothing of importance was found in the Second Pyramid. But inside the Third Pyramid, Vyse's efforts paid off. At the end of July, 1837—as we have briefly mentioned earlier—his workmen broke into its "sepulchral chamber," finding there a beautifully decorated but empty stone "sarcophagus" (Fig. 152). Arabic inscriptions on the walls and other evidence suggested this pyramid "to have been much

Fig. 152

frequented," the floor stones of its chambers and passages "worn and glazed over by the constant passing and repassing of a concourse of people."

Yet in this much frequented pyramid, and in spite of the empty stone coffer, Vyse managed to find proof of its builder—a feat equaling the discovery within the Great Pyramid.

In another rectangular chamber which Vyse called "the large apartment," great piles of rubbish were found, along with the telltale scrawled Arabic graffiti. Vyse at once concluded that this chamber "was probably intended for funeral ceremonies, like those at Abou Simbel, Thebes, etc." When the rubbish was cleared out,

> the greater part of the lid of the sarcophagus was found . . . and close to it, fragments of the top of a mummy-case (inscribed with hieroglyphics, and amongst them, with the cartouche of Menkahre) were discovered upon a block of stone, together with part of a skeleton, consisting of ribs and vertebrae, and the bones of the legs and feet enveloped in coarse woollen cloth of a yellow color . . .
>
> More of the board and cloth were afterwards taken out of the rubbish.
>
> It would therefore seem that, as the sarcophagus could not be removed, the wooden case containing the body had been brought into the large apartment for examination.

This, then, was the scenario outlined by Vyse: Centuries earlier, Arabs had broken in the sepulchral chamber. They found the sarcophagus and removed its lid. Inside there was a mummy within a wooden coffin—the mummy of the pyramid's builder. The Arabs removed coffin and mummy to the large apartment to examine them, breaking them in the process. Now Vyse found all these remains; and a cartouche on

Fig. 153

a fragment of the mummy-case (Fig. 153) spelled out *"Men-ka-ra"*—the very Mycerinus of Herodotus. He proved the identity of the builders of both pyramids!

The sarcophagus was lost at sea during its transportation to England. But the mummy-case and bones reached safely the British Museum, where Samuel Birch could examine the actual inscription, rather than work from facsimiles (as was the case of the inscriptions within the chambers in the Great Pyramid). He soon voiced his doubts: "the coffin of Mycerinus," he said, "manifests considerable difference of style" from monuments of the Fourth Dynasty. Wilkinson, on the other hand, accepted the mummy-case as authentic proof of the identity of the builder of the Third Pyramid; but had doubts regarding the mummy itself: its cloth wrappings did not look to him as being of the claimed antiquity. In 1883, Gaston Maspero concurred "that the wooden coffer-cover of king Menchere is not of the Fourth Dynasty times;" he deemed it a restoration carried out in the Twenty-Fifth Dynasty. In 1892, Kurt Sethe summed up the majority opinion, that the coffin-cover "could have been fashioned only after Twentieth Dynasty times."

As is now well known, both mummy-case and bones were not the remains of an original burial. In the words of I.E.S. Edwards *(The Pyramids of Egypt),* "In the original burial chamber, Col. Vyse had discovered some human bones and the lid of a wooden anthropoid coffin inscribed with the name of Mycerinus. This lid, which is now in the British Museum, cannot have been made in the time of Mycerinus, for it is of a pattern not used before the Saite Period. Radiocarbon tests have shown that the bones date from early Christian times."

The mere statement negating the authenticity of the find does not, however, go to the core of the matter. If the remains were not of an original burial, then they must have been of an intrusive burial; but in such a case, mummy and coffin would be of the same period. This was not the case: here, someone had put together a mummy unearthed in one place, a coffin from another place. The unavoidable conclusion is that the find represented a *deliberate archaeological fraud.*

Could the mismatching have been a coincidence—the genuine remains within the pyramid of *two* intrusive burials, from different times? This must be doubted, in view of the fact that the coffin fragment bore the cartouche of Men-ka-ra. This cartouche has been found on statues and in inscriptions all around the Third Pyramid and its temples (but not inside it), and it is probable that the coffin bearing the cartouche was also found in those surroundings. The coffin's attribution to later times stems not only from its pattern, but also from the wording of the inscription: it is a prayer to Osiris from the Book of the Dead; its appearance on a Fourth Dynasty coffin has been termed remarkable even by the trusting (yet knowledgeable) Samuel Birch *(Ancient History from the Monuments).* Yet it need not have been "a restoration," as some scholars have suggested, from the Twenty-sixth Dynasty. We know from the King List of Seti I from Abydos that the eighth Pharaoh of the Sixth Dynasty was also called Men-ka-ra, and spelled his name in a similar manner.

It is clear, then, that someone had first found, in the vicinity of the pyramid, the coffin. Its importance was surely realized, for—as Vyse himself has reported—he had found just a month previously the name of Men-ka-ra (Mycerinus) written in red paint on the roof of the burial chamber of the middle one of the three small pyramids, south of the Third Pyramid. It must have been this find that gave the team the idea of creating a discovery within the Third Pyramid itself. . . .

The credit for the discovery has been claimed by Vyse and Perring. How could they have perpetrated the fraud, with or without the help of Mr. Hill?

Once again, Vyse's own chronicles hint at the truth. "Not being present when they (the relics) were found," Col. Vyse wrote, he "requested Mr. Raven, when that gentleman was in England, to write an account of the discovery" as an independent witness. Somehow invited to be present at the right moment, Mr. H. Raven, who addressed Col. Vyse as "Sir" and signed the Letter of Evidence "your most obedient servant," attested as follows:

> In clearing the rubbish out of the large entrance-room, after the men had been employed there several days and had advanced some distance towards the south-eastern corner, some bones were first discovered at the bottom of the rubbish; and the remaining bones and parts of the coffin were immediately discovered altogether: no other parts of the coffin or bones could be found in the room.
>
> I therefore had the rubbish, which had been previously turned out of the same room, carefully re-examined, when several pieces of the coffin and of the mummy-cloth were found; but
>
> In no other part of the pyramid were any parts of it to be discovered, although every place was most

minutely examined to make the coffin as complete as possible.

We now get a better grasp of what had happened. For several days, workmen were clearing rubbish from the Large Apartment, piling it up nearby. Though carefully examined, nothing was found. Then, on the last day, as only the southeastern corner of the room remained to be cleared, some bones and fragments of a wooden coffin were discovered. "No other parts of the coffin or bones could be discovered in the room." It was then wisely suggested that the rubbish which had been turned out of the room—a three-foot-high pile—be "carefully re-examined"—not examined, but RE-examined; and—lo and behold—more bones, and coffin fragments with the all-important cartouche, were found!

Where were the remaining parts of the skeleton and coffin? "Although every place was minutely examined to make the coffin as complete as possible, nothing was found in any other part of the pyramid. So, unless we are to believe that bones and coffin fragments were hauled away as souvenirs in centuries past, we can only assume that whoever *hauled in* the discovered parts, brought in just enough fragments to create the discovery: a complete coffin and a complete mummy were either unavailable, or too cumbersome to be smuggled in. ———

Hailed for this second major discovery—he was soon thereafter promoted to the rank of general—Col. Vyse and Perring proceeded to produce at the site of Zoser's step pyramid a stone bearing Zoser's name—written in red paint, of course. There is not enough detail in the chronicles to ascertain whether or not that too was a forgery; but it is indeed incredible that it was again the same tear) who managed to unearth proof of yet one more pyramid builder.

(While most Egyptologists have accepted without further investigation the claim that Khufu's name was inscribed in the Great Pyramid, the works of Sir Alan Gardiner suggest that he had doubts on the subject. In his *Egypt of the Pha-*

raohs, he reproduced royal cartouches with a clear distinction between the hieroglyphs for *Ra* and *Kh.* The cartouche of Cheops, he wrote, "is found in various quarries, in the tombs of his kinfolk and nobles, and in certain writing of later date." Conspicuous by its absence in this list is the inscription in the Great Pyramid . . . Also omitted by Sir Alan were any mention of Vyse's discoveries in the Third Pyramid and even of Vyse's name as such).

If the proof of the construction of the Giza pyramids by the presumed Pharaohs stands shattered, there is no longer reason to suspect the authenticity of the Inventory Stela, which stated that the pyramids and the Sphinx were already there when Khufu came to pay homage to Isis and Osiris.

There is nothing left to contradict our contention that these three pyramids were built by the "gods." On the contrary: everything about them suggests that they were not conceived by men for men's use.

We shall now proceed to show how they were part of the Guidance Grid that served the Spaceport of the Nefilim.

XIV

The Gaze of the Sphinx

In time, the pyramids of Giza were made part of the Landing Grid which had the peaks of Ararat as its focal point, incorporated Jerusalem as a Mission Control Center, and guided the space vehicles to the Spaceport in the Sinai peninsula.

But at first, the pyramids themselves had to serve as guiding beacons, simply by virtue of their location, alignment and shape. All pyramids, as we have seen, were at their core step pyramids—emulating the ziggurats of Mesopotamia. But when the "gods who came from heaven" experimented with their scale model at Giza (the Third Pyramid), they may have found that the silhouette of the ziggurat and the shadow it cast upon the undulating rocks and ever-shifting sands were too blurred and inaccurate to serve as a reliable Pointer-of-the-Way. By casing the stepped core to achieve a "true" pyramid, and using white (light-reflecting) limestone for the casing, a perfect play of light and shadow was achieved, providing clear orientation.

In 1882, as Robert Ballard was watching the Giza pyramids from his train window, he realized that one could determine his location and direction by the ever-changing alignment between the pyramids (Fig. 154). Enlarging on this observation in *The Solution of the Pyramid Problem,* he also showed that the pyramids were aligned with each other in the basic Pythagorean right-angled triangles, whose sides were proportionate to each other as 3:4:5. Pyramidologists have also noticed that the shadows cast by the pyramids could serve as a giant sundial, the direction and length of the shadows indicating time of year and of day.

Even more important, however, was how the silhouettes and shadows of the pyramids appeared to an observer from the skies. As this aerial photograph shows (Fig. 155), the true shape of the pyramids casts arrow-like shadows, which serve as unmistakable direction pointers.

When all was ready to establish a proper Spaceport, it required a much longer Landing Corridor than the one which served Baalbek. For their previous Spaceport in Mesopotamia, the Anunnaki (the biblical Nefilim) chose the most conspicuous mountain in the Near East—Mount Ararat—as their focal point. It should not be surprising that out of the same considerations they again selected it as the focal point of their new Spaceport.

Just as more "coincidences" of triangulation and geometrical perfection have been discovered in the construction and alignment of the Giza pyramids the more they have been examined and studied, so do we find endless "coincidences" of triangulation and alignment as we uncover the Landing Grid laid out by the Anunnaki. If the peaks of Ararat served as the focal point of the new Landing Corridor, then not only the northwestern line of the Landing Corridor, but also its southeastern outline had to be focused on Ararat. But where was its other, Sinai end, anchored on?

Mount St. Katherine lies amidst a massive core of similar, though somewhat lower, granite peaks. When the British Ordnance Survey Mission headed by the Palmers set out to survey the Sinai peninsula, they found that St. Katherine, even if the highest peak, did not stand out sufficiently to serve as a geodesic landmark. Instead, the Mission selected *Mount Umm Shumar* (Fig. 156), which at 8,534 feet is almost a twin in height of Mount St. Katherine (indeed, until the Ordnance Survey, many believed that Umm Shumar was the higher peak). Unlike Katherine, Umm Shumar stands by itself, distinct and unmistakable. Both gulfs can be seen from its peak; its view to the west, northwest, southwest and

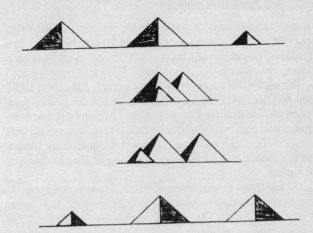

Fig. 154

Fig. 155

east is unobstructed. It was for these reasons that the Palmers without hesitation selected Mount Umm Shumar as their geodesic landmark, the focal point for surveying and measuring the peninsula.

Mount Katherine may have been suitable for a short Landing Corridor focused on Baalbek; but for the distant focal point of Ararat, a much more distinct and unmistakable landmark was required. We believe that for the same reasons as the Palmers', the Anunnaki selected Mount Umm Shumar as the anchor of the southeastern outline of the new Landing Corridor.

Much about this mount and its location is intriguing. To begin with, its name—puzzling or highly significant—means "Mother of Sumer." It is a title which was applied at Ur to Ningal, spouse of Sin. . . .

Unlike Mount St. Katherine, which lies at the center of Sinai's core of high granite peaks and can therefore be reached only with great difficulty, Mount Umm Shumar is situated at the edge of the mass of granite. The sandy beaches there, on the Gulf of Suez, have several natural hot springs. Was it

Fig. 156

there that Asherah spent her winters, residing "by the sea?" From there, it is really only "a she-ass' ride" away to Mount Umm Shumar—a ride so vividly described in the Ugaritic texts when Asherah went calling on El at his Mount.

Just a few miles down the coast from the Hot Springs is the peninsula's most important port city on these coasts—the port city of *el-Tor.* The name—another coincidence?—means "The Bull"; it was, as we have seen, an epithet of El ("Bull El," the Ugaritic texts called him). The place has served as Sinai's most important gulf port from earliest times; and we wonder whether it was not the Tilmun-city (as distinct from Tilmun-land) spoken of in Sumerian texts. It could well have been the port which Gilgamesh planned to reach by ship, from where his comrade Enkidu could go to the nearby mines (in which he was doomed to slave for the rest of his life); while he (Gilgamesh) could proceed to the "Landing Place, where the *Shems* are raised."

The peaks of the peninsula's granite core which face the Gulf of Suez bear names that make one stop and wonder. One mount bears the name "Mount of the Blessed Mother"; closer to Mount Umm Shumar, Mount *Teman* ("The Southern") raises its head. The name brings back the verses of Habakuk: *"El* from *Teman* shall come. . . . Covered are the heavens with his halo; His splendor fills the Earth. . . . The *Word* goes before him, sparks emanate from below; *He pauses to measure the Earth. . . ."*

Was the prophet referring to the mount that still bears that very name—*Teman*—the southern neighbor of Mount "Mother of Sumer?" Since there is no other mountain bearing such a name, the identification seems more than plausible.

Does Mount Umm Shumar fit into the Landing Grid and the network of sacred sites developed by the Anunnaki?

We suggest that this mount substituted for Mt. Katherine when the final Landing Corridor was worked out, acting as the anchor for the southeastern line of the Corridor which was focused on Ararat. But if so, where was the complementary anchor for the northwestern line?

It is no coincidence, we suggest, that Heliopolis was built where it was. *It lies on the original Ararat-Baalbek-Giza line.* But it is so located, that *it is equidistant from Ararat as Umm Shumar is!* Its location was determined, we suggest, by measuring off the distance from Ararat to Umm Shumar—then marking off an equidistant point on the Ararat-Baalbek-Giza line (Fig. 157).

As we unfold the amazing network of natural and artificial peaks that have been incorporated into the landing and communications grid of the Anunnaki, one must ponder whether they served as guiding beacons by height and shape alone. Were they not also equipped with some kind of guidance instruments?

When the two pairs of narrow conduits from the chambers of the Great Pyramid were first discovered, they were thought to have served to lower food to the Pharaoh's at-

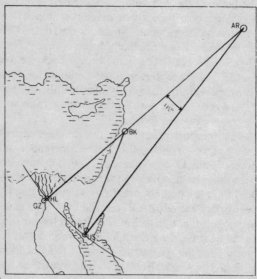

Fig. 157

tendants who were presumed to have been sealed alive in his tomb. When Vyse's team cleared the northern conduit to the "King's Chamber," it was at once filled with cool air; the conduits have since been called "air shafts." This, surprisingly, was challenged by respected scholars in a highly regarded academic publication *(Mitteilungen des Instituts für Orientforschung der Deutschen Akademie der Wissenschaften zu Berlin).* Although the academic establishment has been loath to digress from the "pyramids as tombs" theory, Virginia Trimble and Alexander Badawy concluded in the Bulletin's 1964 issues that the "air shafts" had astronomical functions, having been "beyond doubt inclined within 1° toward the circumpolar stars."

Without doubting that the direction and inclination of the shafts must have been premeditated, we are no less intrigued by the finding that once air flowed into the "King's Chamber", the temperature within it remained at a constant 68° Fahrenheit no matter what the weather outside was. All these findings seem to confirm the conclusions of E. F. Jomard (a member of Napoleon's team of scientists), who suggested that the "King's Chamber" and its "sarcophagus" were not intended for burial, but as a depository of weight and measurement standards, which even in modern times are kept in a stable environment of temperature and moisture.

Jomard could not have possibly imagined—back in 1824—delicate space-guidance instruments, rather than mundane units of a meter and a kilogram. But we, of course, can.

Many who have pondered the purpose of the intricate superstructure of five low compartments above the "King's Chamber," believe they were built to relieve the pressure off the chamber. But this has been achieved in the "Queen's Chamber," with an even greater mass of stone upon it, without such a series of "relieving compartments." When Vyse and his men were inside the compartments, they were astonished to hear clearly every word spoken in other parts of the pyramid. When Flinders Petrie *(The Pyramids and the Temple of Gizeh)* minutely examined the "King's Chamber" and the stone "coffer" within it, he found that both were built in accord with the dimensions of perfect Pythagorian triangles. To cut the coffer out of a solid stone block, he estimated, a saw was needed with nine-foot blades whose teeth were diamond-tipped. To hollow it out, diamond-tipped drills were needed, applied with a pressure of two tons. How all this was achieved was beyond him. And what was the purpose? He lifted the coffer to see whether it hid some aperture (it did not); when the coffer was struck, it emitted a deep, bell-like sound that reverberated throughout the pyramid. This bell-like quality of the coffer was reported by earlier investigators. Were the "King's Chamber" and its "coffer" meant, then, to serve as sound-emitters or echo-chambers?

Even nowadays, landing guidance equipment at airports emits electronic signals which instruments in an approaching aircraft translate into a pleasant buzz if on course; it changes into an alarming beep if the plane veers off course. We can safely assume that, as soon as possible after the Deluge, new guidance equipment was brought down to Earth. The Egyptian depiction of the Divine Cordholders (Fig. 121) indicates that "Stones of Splendor" were installed at both anchor-points of the Landing Corridor; our guess is that the purpose of the various chambers within the pyramid was to house such guidance and communications equipment.

Was *Shad El*—the "Mountain of El"—likewise equipped?

The Ugaritic texts invariably employed the phrase "penetrate the *Shad* of El" when describing the coming of other gods unto the presence of El "within his seven chambers." This implied that these chambers were inside the mountain—as were the chambers inside the artificial mountain of the Great Pyramid.

Historians of the first Christian centuries reported that the people who dwelt in the Sinai and its bordering areas of Palestine and North Arabia worshipped the god *Dushara* ("Lord of the Mountains") and his spouse *Allat,* "Mother of the Gods." They were of course the male El and the female Elat, his spouse Asherah. The sacred object of Dushara was, fortunately, depicted on a coin struck by the Roman governor of those provinces (Fig. 158). Curiously, it resembles the enigmatic chambers within the Great Pyramid—an inclined stairway ("Ascending Gallery") leading to a chamber between massive stones ("The King's Chamber"). Above it, a series of stones re-create the pyramid's "relieving chambers."

Since the Ascending Passages of the Great Pyramid—which are unique to it—were plugged tight when Al Mamoon's men broke into it, the question is: Who, in antiquity, did know, and emulated, the inner construction within the pyramid? The answer can only be: the architects and builders of the Great Pyramid, who possessed such knowledge. Only they could duplicate such construction elsewhere—at Baalbek or within the mountain of El.

Fig. 158

And so it was, that although the Mount of the Exodus was elsewhere, in the northern half of the peninsula, the people of the area transmitted from generation to generation the recollection of sacred mountains among the peninsula's southern peaks. They were the mountains that, by their sheer height and location, and by virtue of the instruments installed within them, served as beacons for the "Riders of the Clouds."

*   *   *

When the first Spaceport was established in Mesopotamia, the flight path was along a center line, drawn precisely in the middle of the arrow-like landing corridor. While guiding beacons flickered their lights and emitted their signals along the two border lines, it was along the central flight path that Mission Control Center was located: the hub of all the communications and guidance equipment, the place where all the computerized information regarding planetary and spacecraft orbits was stored.

When the Anunnaki had landed on Earth and proceeded to establish their facilities and Spaceport in Mesopotamia, Mission Control Center was at Nippur, the "Place of the Crossing." Its "sacred" or Restricted Precinct was under the absolute control of Enlil; it was called the KI.UR ("Earth City"). In its midst, atop an artificially raised platform, was the DUR.AN.KI—"The Bond of Heaven and Earth." It was, the Sumerian texts related, a "heavenward tall pillar reaching to the sky." Firmly set upon the "platform which cannot be overturned," this pillar was used by Enlil "to pronounce the word" heavenward.

That all these terms were Sumerian attempts to describe sophisticated antennas and communications equipment one can gather from the pictographic "spelling" of Enlil's name: it was depicted as a system of large antennas, aerials and a communications structure (see Fig. 52).

Within this "lofty house" of Enlil, there was a hidden, mystery-filled chamber called the DIR.GA—literally meaning "dark, crown-like chamber." Its descriptive name brings so much to mind the hidden, mystifying "King's Chamber" in the Great Pyramid. In the DIR.GA, Enlil and his assistants kept the vital "Tablets of Destinies" on which orbital and space-flight information was stored. When a god who could fly as a bird snatched away these tablets,

> Suspended were the Divine Formulas.
> Stillness spread all over. Silence prevailed . . .
> The sanctuary's brilliance was taken off.

In the DIR.GA Enlil and his assistants kept celestial charts and "carried to perfection" the ME—a term denoting astronaut's instruments and functions. It was a chamber

> As mysterious as distant aeters,
>   as the Heavenly Zenith.
> Among its emblems . . .
>   the emblems of the stars;
> The ME it carries to perfection.
> Its words are for utterance . . .
> Its words are gracious oracles.

A Mission Control Center, similar to the one that had served the landing path in pre-Diluvian Mesopotamia, had to be established for the Spaceport in the Sinai. Where?

Our answer is: in *Jerusalem.*

Hallowed to Jew, Christian and Muslim alike, its very atmosphere charged with some inexplicable unearthly mystery, it had been a sacred city even before King David established her as his capital and Solomon built there the Lord's Abode. When the Patriarch Abraham reached its gates, it was already a well-established center to "*El* the Supreme, the Righteous One of Heaven and Earth." Its earliest known name was *Ur-Shalem*—"City of the Completed Cycle"—a name which suggests an association with orbital matters, or with the God of Orbits. As to who *Shalem* might have been, scholars have offered various theories; some of them (per Benjamin Mazar in "Jerusalem before the David Kingship") name Enlil's grandson Shamash; others prefer Enlil's son Ninib. In all theories, however, the association of Jerusalem's roots with the Mesopotamian pantheon is undisputed.

From its beginnings, Jerusalem encompassed three mountain peaks; from north to south, they were Mount *Zophim,* Mount *Moriah* and Mount Zion. Their names bespoke their functions: The northernmost was the "Mount of Observers" (it is now called in English Mount Scopus); the middle one was the "Mount of Directing"; the southernmost was "The

Mount of the Signal." They are still so called in spite of the passage of millenia.

The valleys of Jerusalem too bear telltale names and epithets. One of them is named in Isaiah the Valley of *Hizzayon,* the "Valley of Vision." The Valley of *Kidron* was known as the "Valley of Fire." In the Valley of *Hinnom* (the *Gehenna* of the Greek New Testament), according to millenia-old legends, there was an entrance to the subterranean world, marked by a column of smoke rising between two palm trees. And the Valley of *Repha'im* was named after the Divine Healers who, according to the Ugaritic texts, were put in the charge of the goddess Shepesh. Aramaic translations of the Old Testament called them "Heroes"; the Old Testament's first translation into Greek called the place the Valley of the *Titans.*

Of the three Mounts of Jerusalem, that of Moriah has been the most sacred. The Book of Genesis explicitly states that it was to one of the peaks of Moriah that the Lord directed Abraham with Isaac, when Abraham's fidelity was tested. Jewish legends relate that Abraham recognized Mount Moriah from a distance, for he saw upon it "a pillar of fire reaching from the earth to heaven, and a heavy cloud in which the Glory of God was seen." This language is almost identical with the biblical description of the descent of the Lord upon Mount Sinai.

The large horizontal platform atop Mount Moriah—reminiscent in layout of the one at Baalbek, though much smaller—has been called "The Temple Mount," for it had served as the site of the Jewish Temple of Jerusalem (Fig. 159). It is now occupied by several Muslim shrines, the most renowned of which is the Dome of the Rock. The dome was carried off by the Caliph Abd al-Malik (seventh century A.D.) from Baalbek, where it adorned a Byzantine shrine; it was erected by the caliph as a roofing over an eight-sided structure he had built to encompass the Sacred Rock: a huge rock to which divine and magical faculties have been attributed from time immemorial.

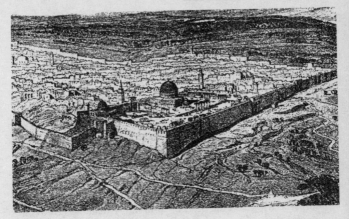

Fig. 159

Muslims believe that it was from the Sacred Rock that their prophet Muhammed was taken aloft to visit Heaven. According to the Koran, Muhammed was taken by the angel Gabriel from Mecca to Jerusalem, with a stopover at Mount Sinai. Then he was taken aloft by the angel, ascending heavenward via a "Ladder of Light." Passing through the Seven Heavens, Muhammed at last stood in the presence of God. After receiving divine instructions, he was brought back to Earth via the same beam of light, landing back at the Sacred Rock. He returned to Mecca, again with a stopover at Mount Sinai, riding the angel's winged horse.

Travelers in the Middle Ages suggested that the Sacred Rock was a huge artificially-cut cube-like rock whose corners faced precisely the four points of the compass. Nowadays, only the top outcropping of the rock can be seen; the presumption of its hidden great cube-like shape may have stemmed from the Muslim tradition that the hallowed Great Stone of Mecca, the *Qa'aba,* was fashioned (on divine instructions) after the Sacred Rock of Jerusalem.

From the visible portion, it is evident that the Sacred Rock had been cut out in various ways on its face and sides, bored through to provide two tube-like funnels, and hollowed out to create a subterranean tunnel and secret chambers. No one knows the purpose of these works; no one knows who had masterminded them and carried them out.

We do know, however, that the First Temple was built by King Solomon upon Mount Moriah at an exact spot and following precise instructions provided by the Lord. The Holy-of-Holies was built upon the Sacred Rock; its innermost chamber, completely gilded, was taken up by two large Cherubim (winged Sphinx-like beings) also made of gold, their wings touching the walls and each other's; between them was placed the Ark of the Testament, from within which the Lord addressed Moses in the desert. Completely insulated from the outside, the gold covered Holy-of-Holies was called in the Old Testament the *Dvir*—literally, "The Speaker."

The suggestion that Jerusalem was a "divine" communication center, a place where a "Stone of Splendor" was secreted, and from which the Word or Voice of the Lord was beamed far and wide, is not as preposterous as it may sound. The notion of such communication was not at all alien to the Old Testament. In fact, the possession by the Lord of such a capability, and the selection of Jerusalem as the communications center, were considered to be attestations of Yahweh's and Jerusalem's supremacy.

"I shall answer the Heavens, and they shall respond to Earth," the Lord assured the Prophet Hosea. Amos prophesied that "Yahweh from Zion will roar; from Jerusalem His voice shall be uttered." And the Psalmist stated that when the Lord shall speak out of Zion, His pronouncements will be heard from one end of Earth to another, and in Heaven too:

> Unto the gods Yahweh hath spoken,
> And the Earth He had called
>   from the east to the west . . .

> The Heavens above He will call,
> and unto the Earth.

Ba'al, the Lord of the facilities at Baalbek, had boasted that his voice could be heard at Kadesh, the gateway city to the Precinct of the gods in the "Wilderness" of central Sinai. Psalm 29, listing some of the places on Earth reachable by the Voice of the Lord of Zion, included both Kadesh and "the cedar place" (Baalbek):

> The voice of the Lord is upon the waters . . .
> The voice of the Lord the cedars breaks . . .
> The voice of the Lord in the Wilderness shall resonance:
> Yahweh the Wilderness of Kadesh shall make shudder.

The capabilities acquired by Ba'al when he installed the "Stones of Splendor" at Baalbek were described in the Ugaritic texts as the ability to put "one lip to Earth, one lip to Heaven." The symbol for these communication devices, as we have seen, were the doves. Both symbolism and terminology are incorporated in the verses of Psalm 68, which describe the flying arrival of the Lord:

> Sing unto the Lord, chant unto His *Shem*,
> Make way for the Rider of the Clouds . . .
> The Lord the Word will utter,
>   the Oracles of an army vast.
> Kings of armies shall escape and flee;
> Abode and home thou shalt divide as spoil—
> Even if they lie between the two Lips and
>   the Dove whose wings are overlaid with silver,
>   whose pinions are greenish gold . . .
> The Chariot of the Lord is mighty,
>   it is of thousands of years;
> Within it the Lord did come
>   from sacred Sinai.

The Jerusalem Stone of Splendor—a "testament stone" or "probing stone" in the words of the Prophets—was secreted in a subterranean chamber. This we learn from a lamentation over the desolation of Jerusalem, when the Lord was wroth with its people:

> The palace was abandoned by the townspeople;
> Forsaken is the peak of Mount Zion (and)
>   the "Prober Which Witnesses."
> The Cavern of Eternal Witnessing
>   is the frolicking place of wild asses,
>   a grazing place of flocks.

With the restoration of the Temple in Jerusalem, the Prophets promised, "the word of Yahweh from Jerusalem shall issue." Jerusalem would be reestablished as a world center, sought by all the nations. Conveying the Lord's promise, Isaiah reassured the people that not only the "probing stone," but also the "measuring" functions would be restored:

> Behold,
> I shall firmly set a Stone in Zion,
>   a Probing Stone,
>   a rare and lofty Stone of Corners,
>   its foundation (firmly) founded.
> He who hath faith,
>   shall not remain unanswered.
> Justice shall be my Cord;
> Righteousness (shall be) my Measure.

To have served as a Mission Control Center, Jerusalem—as Nippur—had to be located on the long central line bisecting the Landing Corridor. Its hallowed traditions affirm such a position, and the evidence suggests that it was that sacred rock which marked the precise geodesic center.

Jerusalem was held by Jewish traditions to have been the "Navel of the Earth." The Prophet Ezekiel referred to the people of Israel as "residing upon the Navel of the Earth"; the Book of Judges related an incident when people were coming down the mountains from the direction of the "Navel of the Earth." The term, as we have seen, meant that Jerusalem was a focal communications center, from which "cords" were drawn to other anchor points of the Landing Grid. It was thus no coincidence that the Hebrew word for the sacred rock was *Eben Sheti'yah*—a term which Jewish sages held to have meant "stone from which the world was woven." The term *sheti* is indeed a weaving term, standing for the long cord that runs lengthwise in a loom (the warp, which is crossed with the shorter weft). It was a fitting term for a stone that marked the exact spot from which the Divine Cords covered Earth as a web.

But as suggestive as all these terms and legends are, the decisive question is this: did Jerusalem in fact lie on the central line which bisected the Landing Corridor, focused on Ararat and outlined by the Giza pyramids and Mount Umm Shumar?

The decisive answer is: Yes. *Jerusalem lies precisely on that line!*

As was the case with the pyramids of Giza, so do we uncover in the case of the Divine Grid more and more amazing alignments and triangulations.

Jerusalem, we find, also *lies precisely where the Baalbek-Katherine line intersects the flight path's central line based on Ararat.*

Heliopolis, we find, is *precisely equidistant from Jerusalem as Mount Umm Shumar.*

And the diagonals drawn from Jerusalem to Heliopolis and to Umm Shumar form *an accurate 45° angle* (Fig. 160)!

These links between Jerusalem, Baalbek (The Crest of Zaphon) and Giza (Memphis) were known, and hailed, in biblical times:

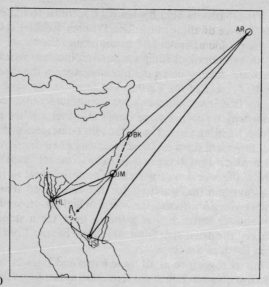

Fig. 160

> Great is Yahweh and greatly hallowed
>   in the city of our Lord,
> His Holy Mountain.
> At Memphis He is beautified.
> The joy of the whole Earth,
>   of Mount Zion,
>   of the Crest of Zaphon.

Jerusalem, the Book of Jubilees held, was in fact one of four "Places of the Lord" on Earth: "the Garden of Eternity" in the Cedar Mountain; "the Mountain of the East" which was Mount Ararat; Mount Sinai and Mount Zion. Three of them were in the "lands of Shem," the son of Noah from whom the biblical Patriarchs were descended; and they were interconnected:

The Garden of Eternity, the most sacred,
   is the dwelling of the Lord;
And Mount Sinai, in the center of the desert;
And Mount Zion, the center of the Navel of the Earth.
These three were created as holy places,
FACING EACH OTHER.

Somewhere along the "Line of Jerusalem," the central flight line that was anchored on Mount Ararat, the Spaceport itself had to be located. There, too, the final beacon had to be located: "Mount Sinai, *in the center of the desert.*"

It is here, we suggest, that the dividing line which we now call the Thirtieth Parallel (north) had come into play.

We know from Sumerian astronomical texts that the skies enveloping Earth were so divided as to separate the northern "way" (allotted to Enlil) from the southern "way" (allotted to Ea) with a wide central band considered the "Way of Anu." It is only natural to assume that a dividing line between the two rival brothers should also have been established after the Deluge, when the settled Earth was divided into the Four Regions; and that, as in pre-Diluvial times, the Thirtieth Parallels (north and south) served as demarcation lines.

Was it mere coincidence, or a deliberate compromise between the two brothers and their feuding descendants, that in each of the three regions given to Mankind, the sacred city was located on the Thirtieth Parallel?

The Sumerian texts state that "When Kingship was lowered from Heaven" after the Deluge, "Kingship was in Eridu." Eridu was situated astride the Thirtieth Parallel as close to it as the marshy waters of the Persian Gulf had permitted. While the administrative-secular center of Sumer shifted from time to time, Eridu remained a sacred city for all time.

In the Second Region (the Nile Civilization) the secular capital also shifted from time to time. But Heliopolis forever remained the sacred city. The Pyramid Texts recognized its links with other sites, and called its ancient gods "Lords of the Dual Shrines." These two paired shrines bore the intriguing (and pre-Egyptian?) names *Per-Neter* ("Coming-forth Place of the Guardians") and *Per-Ur* ("The Coming-forth Place of Old"); their hieroglyphic depictions bespoke great antiquity.

The dual or paired shrines played a major role in the Pharaonic succession. During the rites, conducted by the *Shem* priest, the crowning of the new king and his admission to the "Place of the Guardians" in Heliopolis coincided with the departure of the deceased king's spirit, through the eastern False Door, to the "Coming-forth Place of Old."

And Heliopolis was located astride the Thirtieth Parallel, as close to it as the Nile's delta permitted!

When the Third Region, the Indus Valley Civilization, followed, its secular center was on the shores of the Indian Ocean; but its sacred city—*Harappa*—was hundreds of miles away to the north—right on the Thirtieth Parallel.

The imperative of the northern Thirtieth Parallel appears to have continued in the millenia that followed. Circa 600 B.C., the Persian kings augmented the royal capital with a city "Sacred unto all Nations." The place selected for its construction was a remote and uninhabited site. There, literally in the middle of nowhere, a great horizontal platform was laid out. Upon it, palaces with magnificent staircases and many auxiliary shrines and structures were erected—all honoring the God of the Winged Globe (Fig. 161). The Greeks called the place *Persepolis* ("City of the Persians"). No people lived there: It was only to celebrate the New Year on the day of the spring equinox that the king and his retinue came there. Its remains still stagger the viewer. And it was located astride the Thirtieth Parallel.

No one knows for sure when *Lhasa* in Tibet—the sacred city of Buddhism—was founded. But it is a fact that Lhasa too—as Eridu, Heliopolis, Harappa and Persepolis were—was situated on the same Thirtieth Parallel (Fig. 162).

The sanctity of the Thirtieth Parallel must be traced back

Fig. 161

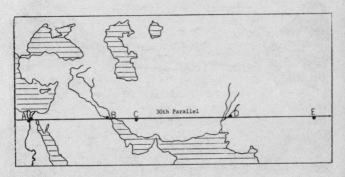

(A) Giza-Heliopolis (B) Eridu (C) Persepolis (D) Harappa (E) Lhasa

Fig. 162

to the origins of the Sacred Grid, when the divine measurers determined the location of the pyramids of Giza also on the Thirtieth Parallel. Could the gods have given up this "sanctity" or neutrality of the Thirtieth Parallel when it came to their most vital installation—the Spaceport—in their own Fourth Region, in the Sinai peninsula?

It is here that we ought to seek a final clue from the remaining enigma of Giza—its Great Sphinx. Its body is that of a crouching lion, its head of a man wearing the royal head-dress (Fig. 163). When and by whom was it erected? And to what purpose? Whose image does it bear? And why is it where it is, alone, and nowhere else?

The questions have been many, the answers very few. But

Fig. 163

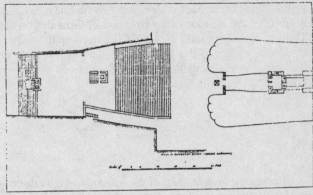

Fig. 164

one thing is certain: *it gazes precisely eastward, along the Thirtieth Parallel.*

This precise alignment and gaze eastward along the Divine Parallel were emphasized in antiquity by a series of structures that extended from in front of the Sphinx eastward precisely along an east-west axis (Fig. 164).

When Napoleon and his men saw the Sphinx at the turn of the eighteenth century, only its head and shoulders protruded above the desert sands; it was in that state that the Sphinx was depicted and known for the better part of the century that followed. It took repeated and systematic excavations to reveal its full colossal size (240 feet long, 65 feet high) and shape, and to confirm what ancient historians had written: that it was a single piece of sculpture, carved by some giant hand out of the natural rock. It was none other than Capt. Caviglia, whom Col. Vyse forced out of Giza, who had uncovered during 1816–1818 not only a good part of the body and extended paws of the Sphinx, but also the temples, sanctuaries, altars and stelas that were erected in front of it.

Fig. 165

Clearing the area in front of the Sphinx, Caviglia discovered a platform that extended somewhat on both sides of the Sphinx but which primarily ran eastward. Excavating for one hundred feet in that easterly direction, he came upon a spectacular staircase of thirty steps leading up to a landing; upon it were the remains of what looked like a pulpit. At the eastern end of the landing, some forty feet away, another flight of thirteen steps was discovered; they raised the level to the same height as the head of the Sphinx.

There, a structure whose function was to support two columns (Fig. 165) was so situated that the eastward gaze of the Sphinx passed precisely between the two columns.

Archaeologists believe that these remains are from Roman times. But as we have seen at Baalbek, the Romans embellished monuments that predated their era, building and rebuilding where earlier monuments and shrines had stood. It is well established by now, that Greek conquerors and Roman emperors continued the tradition of the Pharaohs to visit and pay homage to the Sphinx, leaving behind appropriate inscriptions. They affirmed the belief, which continued into Arab times, that the Sphinx was the work of the gods themselves; it was deemed to be the harbinger of a future messianic era of peace. An inscription by the notorious emperor Nero called the Sphinx *"Armachis, Overseer and Savior."*

Because the Great Sphinx is situated near the causeway leading to the Second Pyramid, the best idea scholars had to offer was that it was built by Chefra, the "builder" of the Second Pyramid; and that it must therefore bear his image. This notion is devoid of any factual basis; yet it has persisted in textbooks, although as far back as 1904 E. A. Wallis Budge, then Keeper of the Egyptian and Assyrian Antiquities in the British Museum, concluded unequivocally *(The Gods of the Egyptians)* that "this marvellous object was in existence in the days of Kha-f-ra, or Khephren; and it is probable that it is a very great deal older than his reign and that it dates from the end of the archaic period."

As the "Inventory Stela" attests, the Sphinx had already stood at Giza in the time of Khufu, a predecessor of Chefra. Like several Pharaohs after him, so did Khufu take credit for removing the sand that encroached upon the Sphinx. From this it must be deduced that the Sphinx was already an olden monument in Khufu's time. What earlier Pharaoh, then, had erected it, implanting upon it his own image?

The answer is that the image is not of any Pharaoh, but of a god; and that in all probability, the gods and not a mortal king had erected the Sphinx.

Indeed, only by ignoring what the ancient inscriptions had stated, could anyone assume otherwise. A Roman inscription, calling the Sphinx "Sacred Guide," said of it: "Thy formidable form is the work of the Immortal Gods." A Greek adolatory poem read in part:

> Thy formidable form,
> Here the Immortal Gods have shaped. . . .
> As a neighbor to the Pyramids they placed thee. . . .
> A heavenly monarch who his foes defies. . . .
> Sacred Guide in the Land of Egypt.

In the Inventory Stela, Khufu called the Sphinx "Guardian of the Aeter, who guides the Winds with his gaze." It was, as he clearly wrote, the image of a god:

> This figure of the god
>> will exist to eternity;
> Always having its face
>> watching towards the east.

In his inscription, Khufu mentions that a very old syca-more tree that grew near the Sphinx was damaged "When the Lord of Heaven descended upon the Place of *Hor-em-Akhet,*" "the Falcon-god of the Horizon." This, indeed was the most frequent name of the Sphinx in Pharaonic inscriptions; his other epithets being *Ruti* ("The Lion") and *Hul* (meaning, perhaps, "The Eternal").

Nineteenth century excavators of the site of the Sphinx, records show, were prompted by local Arab lore that held that there existed, under or within the Sphinx, secret chambers holding ancient treasures or magical objects. Caviglia, as we have seen, exerted himself within the Great Pyramid in search of a "hidden chamber;" it appears that he had switched to the pyramid having failed to find such a chamber at the site of the Sphinx. Perring too made the attempt, by cutting forcibly a deep hole in the back of the Sphinx.

Even more responsible researchers, such as Auguste Mariette in 1853, shared the general opinion that there is a hidden chamber concealed in or under the Sphinx. This belief was bolstered by the writings of the Roman historian Pliny, who reported that the Sphinx "contained the tomb of a ruler named Harmakhis," and by the fact that nearly all ancient depictions of the Sphinx show it crouching atop a stone structure. The searchers have surmised that if the Sphinx itself could have been almost hidden from sight by the encroaching sands, so much so could the sands of desert and time completely hide any substructure.

The most ancient inscriptions seem to suggest that there indeed existed not one, but two secret chambers under the Sphinx—perhaps reachable through an entrance hidden under the paws of the monument. A hymn from the time of the Eighteenth Dynasty, moreover, reveals that the two "caverns" under the Sphinx enabled it to serve as a communications center!

The god Amen, the inscription said, assuming the functions of the heavenly Hor-Akhti, attained "perception in his heart, command on his lips . . . when he enters the two caverns which are under his [the Sphinx's] feet." Then,

> A message is sent from heaven;
> It is heard in Heliopolis,
>   and is repeated in Memphis by the Fair of Face.
> It is composed in a despatch by the writing of Thoth,
>   with regard to the city of Amen (Thebes) . . .
> The matter is answered in Thebes,
> A statement is issued . . . a message is sent.
> The gods are acting according to command.

In the days of the Pharaohs it was believed that the Sphinx—though sculpted out of stone—could somehow hear and speak. In a long inscription on a stela (Fig. 166) erected between the paws of the Sphinx by Thothmes IV (and dedicated to the emblem of the Winged Disk), the king related that the Sphinx spoke to him and promised him a long and prosperous reign if only he would remove the sands that encroached upon his (the Sphinx's) limbs. One day, Thothmes wrote, as he went hunting out of Memphis, he found himself on the "sacred road of the gods which led from Heliopolis to Giza. Tired, he lay to rest in the shade of the Sphinx; the place, the inscription reveals, was called *"Splendid Place of the Beginning of Time."* As he fell asleep by this "very great statue of the Creator," the Sphinx—this "majesty of the Revered God"—began to speak to him, introducing itself by saying: "I am thy ancestor *Hor-em-Akhet,* the one created of Ra-Aten."

Many unusual "Ear Tablets" and depictions of the Twin Doves—a symbol associated with oracle sites—were found in the temples surrounding the Sphinx. Like the ancient inscriptions, they too attest to the belief that somehow the Sphinx could transmit Divine Messages. Although the efforts to dig under the Sphinx have not been successful, one cannot rule out the possibility that the subterranean chambers which the gods had entered with "command on their lips" would still be found.

It is clear from numerous funerary texts that the Sphinx was considered to have been the "Sacred Guide" who guided the deceased from "yesterday" to "tomorrow." Coffin Spells intended to enable the deceased's journey along the "Path of the Hidden Doors" indicate that it began at the site of the Sphinx. Invoking the Sphinx, the Spells asserted that "The Lord of Earth has commanded, the Double Sphinx has repeated." The journey began when *Hor-Akhet*—the Sphinx—

Fig. 166

pronounced: "Pass by!" Drawings in the *Book of the Two Ways,* which illustrated the journey, show that from the starting point at Giza there were two routes by which the *Duat* could be reached.

As the Sacred Guide, the Sphinx was often depicted guiding the Celestial Barge. Sometimes, as on the stela of Thothmes (Fig. 166) it was depicted as a double Sphinx, guiding the Celestial Barge from "yesterday" to "tomorrow." In this role it was associated with the Hidden God of the Subterranean Realm; it was as such, it will be recalled (Fig. 19) that it symbolically appeared flanking the hermetically sealed chamber of the god Seker in the *Duat.*

Indeed, the Pyramid Texts and the Book of the Dead refer to the Sphinx as "The Great God who opens the Gates of Earth"—a phrase that may suggest that the Sphinx at Giza, which "led the way," had a counterpart near the Stairway to Heaven, who opened there "the Gates of Earth." Such a possibility is perhaps the only explanation (in the absence of any other to date) of a very archaic depiction of the Pharaoh's Journey to the Afterlife (Fig. 167). It begins with a crouching Horus-symbol which gazes toward the Land of the Date Palm where an unusual vessel with dredges or

Fig. 167

cranes (?) is situated, as well as a structure which is reminiscent of the Sumerian depiction of the name EN.LIL as a communications center (Fig. 52). A god greeting the Pharaoh, a bull and the Bird of Immortality are seen, followed by fortifications and an assortment of symbols. Finally, the symbol for "place" (tilted cross within a circle) appears between the sign for the Stairway and *a sphinx looking the other way!*

A stela erected by one Pa-Ra-Emheb, who directed works of restoration at the site of the Sphinx in Pharaonic times, contains telltale verses in adoration of the Sphinx; their similarity to biblical Psalms is truly tantalizing. The inscription mentions the extension of cords "for the plan," the making of "secret things" in the subterranean realm; they speak of the "crossing of the sky" in a Celestial Barge, and of a "protected place" in the "sacred desert." It even employs the term *Sheti.ta* to denote the "Place of the Hidden Name" in the Sacred Desert:

> Hail to thee, King of the Gods,
> Aten, Creator . . .
> Thou extendest the cords for the plan,
>   thou didst form the lands . . .
> Thou didst make secret the Underworld . . .
> The Earth is under thy leading;
>   thou didst make high the sky . . .
> Thou hast built for thee a place protected
>   in the sacred desert, with hidden name.
> Thou risest by day opposite them . . .
> Thou art rising beautifully . . .
> Thou art crossing the sky with a good wind . . .
> Thou art traversing the sky in the barque . . .
> The sky is jubilating,
> The Earth is shouting of joy.
> The crew of Ra do praising every day;
> He comes forth in triumph.

To the Hebrew Prophets, the *Sheti*—the central Flight Line passing through Jerusalem—was the Divine Line, the direction to watch: "within it did the Lord come from sacred Sinai."

But to the Egyptians, as the above inscription declared, *Sheti.ta* was the "Place of the Hidden Name." It was in the "Sacred Desert"—which is exactly what the biblical term "Desert of *Kadesh*" has meant. And to it, the "cords of the plan" were extended from the Sphinx. There, Paraemheb had seen the King of the Gods ascend by day; the words are almost identical to those of Gilgamesh, arriving at Mount Mashu, "where daily the *Shems* he watched, as they depart and come in . . . watched over Shamash as he ascends and descends."

It was the Protected Place, the Place of Ascent. Those who were to reach it were guided there by the Sphinx; for its gaze led eastward, exactly along the Thirtieth Parallel.

It was where the two lines intersected, we suggest—where the Line of Jerusalem intersected the Thirtieth Parallel that the Gates of Heaven and Earth were located: the Spaceport of the gods.

The intersection is located within the Sinai's Central Plain, As the *Duat* was depicted in the Book of the Dead, the Central Plain is indeed an oval plain encompassed by mountains. It is a vast valley whose surrounding mountains are separated by seven passes—as described in the *Book of Enoch;* a vast flat plain whose hard natural surface provided ready-made runways for the shuttlecraft of the Anunnaki.

Nippur, we have shown (see Fig. 122), was the focal point, the bulls-eye of concentric circles which measured off as equidistant the Spaceport at Sippar and other vital installations and sites. The same, we find not without amazement, held true for Jerusalem (Fig. 168):

• The Spaceport (SP) and the Landing Place at Baalbek (BK) lay on the perimeter of an inner circle,

forming a vital team of installations that were equidistant from the Control Center in Jerusalem (JM);
• The geodesic beacon of Umm Shumar (US) and the beacon of Heliopolis (HL) lay on the perimeter of an outer circle, making them too a pair equidistant from Jerusalem.

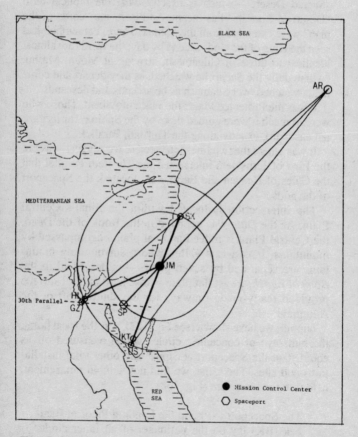

Fig. 168

As we fill in our chart, the masterful Grid conceived by the Anunnaki unfolds before our very eyes; and we are truly astounded by its precision, simple beauty, and the artful combination of basic geometry with the landmarks provided by nature:

- The Baalbek-Katherine line, and the Jerusalem-Heliopolis line, intersected each other at the basic and precise angle of 45°; the central flight path bisected this angle into two precise angles of 22½° each; the grand Flight Corridor was in turn precisely half that (11¼°);
- The Spaceport, situated at the intersection of the central Sight path and the Thirtieth Parallel, was equidistant from Heliopolis and Umm Shumar.

Was it only an accident of geography that Delphi (DL) was equidistant from Mission Control in Jerusalem and from

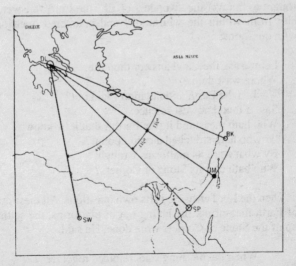

Fig. 169

the Spaceport in central Sinai? Mere coincidence that the angular width of the (Flight?) corridor so created was 11¼°? That another Flight Corridor of 11¼° connected Delphi with Baalbek (BK)?

Or only chance that the lines connecting Delphi with Jerusalem and the oasis of Siwa (SW)—the site of Ammon's oracle to which Alexander had rushed—formed once again the angle of 45° (Fig. 169)?

Were other sacred cities and oracle sites in Egypt, such as the great Thebes and Edfu, located where they were at a king's whim, at an attractive bend of the Nile—or where alignments of the Grid had dictated?

Indeed, were we to study all these sites, all of Earth would probably be encompassed. But was that not what Ba'al had already known when he established his clandestine facilities at Baalbek? For his aim, we recall, was to communicate with and dominate, not just the nearby lands, but all of Earth.

This the biblical Lord too must have known; for when Job sought to unravel the "wonders of El," the Lord "answered him from within the whirlpool," and countered questions with questions:

> Let me ask thee, and answer thou me:
> Where wast thou,
>   when the Earth's foundation I laid out?
> Say, if thou knowest science:
> Who hath measured it (the Earth), that it be known?
> Or who hath stretched a cord upon it?
> By what were its platforms wrought?
> Who hath cast its Stone of Corners?

Then the Lord answered His own questions. All these acts of Earth measuring, the laying out of platforms, the setting up of the Stone of Corners were done, He said:

> When the morning stars rejoiced together
> And all the sons of the gods shouted for joy.

Man, as wise as he might have been, had no hand in all that. Baalbek, the Pyramids, the Spaceport—all were meant for the gods alone.

But Man, ever searching for Immortality, has never ceased to follow the gaze of the Sphinx.

Sources

In addition to works specifically mentioned in the text, the following served as principal sources on the ancient Near East:

I. Studies and articles in various issues of the following periodicals:

Ägyptologische Forschungen (Hamburg-New York).
Der Alte Orient (Leipzig).
American Journal of Archeology (Concord, N.H.).
American Journal of Semitic Languages and Literature (Chicago).
Ametocan Philosophical Society, Memoirs (Philadelphia).
Analecta Orientalia (Rome).
Annales du Musée Guimet (Paris).
Annales du Service des Antiquites de l'Egypte (Cairo).
Annual of the American Schools of Oriental Research (New Haven).
Annual of the Palestine Exploration Fund (London).
Antiquity (Cambridge).
Archaeologia (London).
Archiv für Keilschriftforschung (Berlin).
Archiv für Orientforschung (Berlin).
Archiv Orientālni (Prague).
The Assyrian Dictionary of the Oriental Institute, University of Chicago (Chicago).
Assyriologische Bibliothek (Leipzig).

Assyriological Studies of the Oriental Institute, University of Chicago (Chicago).

Babyloniaca (Paris).
Beiträge zur Aegyptischen Bauforschung und Altertumskunde (Kairo).
Beiträge zur Assyriologie und semitischen Sprachwissenschaft (Leipzig).
Biblical Archaeology Review (Washington).
Bibliotheca Orientalis (Leiden).
British School of Archaeology and Egyptian Research, Account Publications (London).
Bulletin de l'institut français d'archeologie orientale (Cairo).
Bulletin of the American Schools of Oriental Research (New Haven).

Cuneiform Texts from Babylonian Tablets in the British Museum (London).

Deutsche Orient-Geselschaft, Mitteilungen (Berlin).
Deutsche Orient-Geselschaft, Sendschriften (Berlin).

Egypt Exploration Fund, Memoirs (London).
Ex Oriente Lux (Leipzig).

France: Délégation en Perse, Memoires (Paris).
France: Mission Archéologique de Perse, Memoires (Paris).

Harvard Semitic Series (Cambridge, Mass.).
Hispanic American Historical Review (Durham, N.C.).

Iraq (London).
Imperial and Asiatic Quarterly Review (London).
Institut Français d'Archéologie Orientale, Bibliothèque d'Etude (Cairo).

Institut François d'Archéologie Orientale, Memoires
   (Cairo).
Israel Exploration Society, Journal (Jerusalem).

Jewish Palestine Exploration Society, Bulletin (Jerusalem).
Journal of the American Oriental Society (New Haven).
Journal of Biblical Literature and Exegesis (Philadelphia).
Journal of Cuneiform Studies (New Haven and Cambridge,
   Mass.).
Journal of Egyptian Archaeology (London).
Journal of Jewish Studies (Oxford).
Journal of Near Eastern Studies (Chicago).
Journal of the Palestine Oriental Society (Jerusalem).
Journal of the Royal Asiatic Society (London).
Journal of Sacred Literature and Biblical Record
   (London).
Journal of the Society of Oriental Research (Chicago).

Kaiserlich Deutschen Archaelogischen Institut, Jahrbuch
   (Berlin).
*Königliche Akademie der Wissenschaften zu Berlin,
   Abhandlungen* (Berlin).

Leipziger Semitische Studien (Leipzig).

Mitteilungen der altorientalischen Gesellschaft (Leipzig).
*Mitteilungen des deutschen Instituts für ägyptische
   Altertumskunde in Kairo* (Augsburg and Berlin).
Mitteilungen des Instituts für Orientforschung (Berlin).

Orientalia (Rome).
Orientalistische Literaturzeitung (Leipzig).

Palestine Exploration Quarterly (London).
Preussischen Akademie der Wissenschaften, Abhandlungen
   (Berlin).

Proceedings of the Society of Biblical Archaeology (London).

Qadmoniot, Quarterly for the Antiquities of Eretz-Israel and Bible Lands (Jerusalem).

Recueil de travaux relatifs à la philologie et à l'archéologie égyptiennes et assyriennes (Paris).
Revue Archéologique (Paris).
Revue d'Assyriologie et d'archéologie orientale (Paris).
Revue Biblique (Paris).

Sphinx (Leipzig).
Studio Orientalia (Helsinki).
Studies in Ancient Oriental Civilizations (Chicago).
Syria (Paris).

Tarbiz (Jerusalem).
Tel Aviv, Journal of the Tel-Aviv University Institute of Archaeology (Tel-Aviv).
Transactions of the Society of Biblical Archaeology (London).

Untersuchungen zur Geschichte und Altertumskunde Aegyptens (Leipzig).
Urkunden des ägyptischen Altertums (Leipzig).

Vorderasiatisch-Aegyptischen Gesellschaft, Mitteilungen (Leipzig).
Vorderasiatische Bibliothek (Leipzig).

Die Welt des Orients (Göttingen).
Wissenschaftliche Veröffentlichungen der Deutschen Orient-Gesellschaft (Berlin and Leipzig).

Yale Oriental Series, Babylonian Texts (New Haven).

Yerushalayim, Journal of the Jewish Palestine Exploration Society (Jerusalem).

Zeitschrift für ägyptische Sprache und Altertumskunde (Berlin).
Zeitschrift für die alttestamentliche Wissenschaft (Berlin and Giessen).
Zeitschrift für Assyriologie und verwandte Gebiete (Leipzig).
Zeitschrift der Deutsche morgenländische Gesellschaft (Leipzig).
Zeitschrift des deutschen Palaestina-Vereins (Leipzig).
Zeitschrift für Keilschriftforschung und verwandte Gebiete (Leipzig).
Zeitschrift für die Kunde des Morgenlandes (Göttingen).

II. Individual works:

Alouf, M. M.: *History of Baalbek* (1922).
Amiet, P.: *La Glyptique Mésopotamienne Archaique* (1961).
Antoniadi, E. M.: *L'Astronomie Égyptienne* (1934).
Avi-Yonah, M.: *Sefer Yerushalaim* (1956).

Babelon, E.: *Les Rois de Syrie* (1890).
———: *Les Collections de Monnais Anciennes* (1897).
———: *Traité des Monnais Greques et Romaines* (1901–1910).
Bauer, H.: *Die alphabetischen Keilschrifttexte von Ras Schamra* (1936).
Borchardt, L.: *Die Entstehung der Pyramide* (1928).
Bourguet, E.: *Les Ruines de Delphos* (1914).
Buck, A. de: *The Egyptian Coffin Texts* (1935–1961).
Budge, E.A.W.: *The Alexander Book in Ethiopia* (1933).
———: *Cleopatra's Needle* (1906).
———: *The Egyptian Heaven and Hell* (1906).
———: *Egyptian Magic* (1899).

————: *The Gods of the Egyptians* (1904).
————: *The History of Alexander the Great* (1889).
————: *The Life and Exploits of Alexander the Great* (1896).
————: *Osiris and the Egyptian Resurrection* (1911).
Budge, E.A.W. and King, L. W.: *Annals of the Kings of Assyria* (1902).

Capart, J.: *Recueil de Monuments Égyptiens* (1902).
————: *Thebes* (1926).
Cassuto, M. D.: *Ha'Elah Anath* (1951).
————: *Perush al Sefer Shemoth* (1951).
Contenau, G.: *L'Épopée de Gilgamesh* (1939).

Davis, Ch. H. S.: *The Egyptian Book of the Dead* (1894).
Delaporte, L.: *Catalogue des Cylindres Orientaux* (1910).
Delitzsch, F.: *Wo Lag Das Paradies?* (1881).
Dussaud, R.: *Notes de Mythologie Syrienne* (1905).
————: *Les Découvertes de Ras Shamra (Ugarit) et l'Ancien Testament* (1937).

Ebeling, E.: *Reallexikon der Assyriologie* (1928–1932).
Eckenstein, L.: *A History of Sinai* (1921).
Emery, W. B.: *Excavations at Saqqara* (1949–58).
Erman, A.: A *Handbook of Egyptian Religion* (1907).
————: *Aegypten und Aegyptisches Leben im Altertum* (1923).
————: *The Literature of the Ancient Egyptians* (1927).

Falkenstien, A.: *Literarische Keilschrifttexte aus Uruk* (1931).
Faulkner, R. O.: *The Ancient Egyptian Coffin Texts* (1973).
————: *The Ancient Egyptian Pyramid Texts* (1969).
Frankfort, H.: *Kingship and the Gods* (1948).
Frauberger, H.: *Die Akropolis von Baalbek* (1892).
Friedländer, I.: *Die Chadirlegende und der Alexanderroman* (1913).

Gaster, Th. H.: *Myth, Legend and Custom in the Old Testament* (1969).
Gauthier, H.: *Dictionnaire des Noms Geographique* (1925).
Ginsberg, L.: *Kitbe Ugarit* (1936).
———: *The Legends of the Jews* (1954).
———: *The Ras Shamra Mythological Texts* (1958).
Gordon, C. H.: *The Loves and Wars of Baal and Anat* (1943).
———: *Ugaritic Handbook* (1947).
———: *Ugaritic Literature* (1949).
Gray, J.: *The Canaanites* (1965).
Gressmann, E.: *Altorientalische Texte zum alten Testament* (1926).
Grinsell, L. V.: *Egyptian Pyramids* (1947).

Heidel, A.: *The Gilgamesh Epic and Old Testament Parallels* (1946).
Hooke, S. H.: *Middle Eastern Mythology* (1963).
Hrozny, B.: *Hethitische Keilschrifttexte aus Boghazköy* (1919).

Jensen, P.: *Assyrisch-Babylonische Mythen und Epen* (1900).
———: *Das Gilgamesch-Epos in der Weltliteratur* (1906, 1928).
Jéquier, G.: *Le Livre de ce qu'il y a dans l'Hades* (1894).

Kazis, I. J.: *The Book of the Gests of Alexander of Macedon* (1962).
Kees, H.: *Aegyptische Kunst* (1926).
Kenyon, K. M.: *Jerusalem* (1967).
Kraeling, E. G. (Ed.): *Historical Atlas of the Holy Land* (1959).
Kramer, S. N.: *Gilgamesh and the Huluppu Tree* (1938).
———: *Sumerian Mythology* (1944).

Langdon, S.: *Historical and Religious Texts* (1914).
———. *The Epic of Gilgamesh* (1917).

Leonard, W. E.: *Gilgamesh* (1934).
Lefébure, M. E.: *Les Hypogées Royaux de Thébes* (1882).
Lepsius, K. R.: *Auswahl der wichtigsten Urkunden des Aegyptischen Alterthums* (1842).
———: *Königsbuch der Alten Aegypter* (1858).
Lesko, L. H.: *The Ancient Egyptian Book of the Two Ways* (1972).
Lipschitz, O.: *Sinai* (1978).
Luckenbill, D. D.: *Ancient Records of Assyria and Babylonia* (1926–1927).

Meissner, B.: *Alexander und Gilgames* (1894).
Mercer, S.A.B.: *Horus, Royal God of Egypt* (1942).
Meshel, Z.: *Derom Sinai* (1976).
Montet, P.: *Eternal Egypt* (1969).
Montgomery, J. A., and Harris, R. S.: *The Ras Shamra Mythological Texts* (1935).
Müller, C.: *Pseudokallisthenes* (1846).

Naville, H. E.: *Das aegyptische Todtenbuch* (1886).
Nöldeke, Th.: *Beiträge zur Geschichte des Alexanderromans* (1890).
Noth, M.: *Geschichte Israels* (1956).
———: *Exodus* (1962).

Obermann, J.: *Ugaritic Mythology* (1948).
Oppenheim, A. L.: *Mesopotamian Mythology* (1948).

Perlman, M. and Kollek, T.: *Yerushalayim* (1969).
Perring, J. E.: *The Pyramids of Gizeh from Actual Survey and Measurement* (1839).
Petrie, W.M.F.: *The Royal Tombs of the First Dynasty* (1900).
Poebel, A.: *Sumerische Studien* (1921).
Porter, B. and Moss, R.L.B.: *Topographical Bibliography of Ancient Egypt* (1951).
Pritchard, James B: *Ancient Near Eastern Texts Relating to the Old Testament* (3rd ed., 1969).

————: *The Ancient Near East in Pictures Relating to the Old Testament* (1969).

Puchstein, O.: *Führer durch die Ruinen von Baalbek* (1905).

————: *Guide to Baalbek* (1906).

Puchstein, O. and Lupke, Th. von: *Baalbek* (1910).

Rawlinson, H. C.: *The Cuneiform Inscriptions of Western Asia* (1861–1884).

Reisner, G. A.: *Mycerinus: The Temples of the 3rd Pyramid at Gizeh* (1931).

Ringgren, H.: *Israelitische Religion* (1963).

Rothenberg, B. and Aharoni, Y.: *God's Wilderness* (1961).

Rougé, E. de: *Recherches sur le Monuments qu'on peut Attributer aux six premières dynasties de Manethon* (1866).

Schott, A.: *Das Gilgamesch-Epos* (1934).

Schrader, E. (Ed.): *Keilinschriftliche Bibliothek* (1889–1900).

Soden, W. von: *Sumerische und Akkadische Hymnen und Gebete* (1953).

Smyth, C. P.: *Life and Work at the Great Pyramid* (1867).

Thompson, R. C.: *The Epic of Gilgamesh* (1930).

Ungnad, A.: *Die Religion der Babylonier und Assyrer* (1921).

————: *Das Gilgamesch Epos* (1923).

————: *Gilgamesch Epos und Odyssee* (1923).

Ungnad, A. and Gressmann, H.: *Das Gilgamesch-Epos* (1919).

Vandier, J.: *Manuel d'Archéologie Égyptienne* (1952).

Virolleaud, Ch.: *La déesse 'Anat* (1938).

————: *La légende phénicienne de Danel* (1936).

Volney, C. F.: *Travels Through Syria* (1787).

Wainwright, G. A.: *The Sky Religion in Ancient Egypt* (1938).
Weidner, E. F.: *Keilschrifttexte aus Boghazkoy* (1916).
Wiegand, Th.: *Baalbek* (1921–1925).
Woloohjian, A. M.: *The Romance of Alexander the Great by Pseudo-Callisthenes* (1969).

Zimmern, H.: *Sumerische Kultlieder* (1913).

Index

Turn the page
for a revealing sneak preview of

THE
END *of* DAYS

by
Zecharia Sitchin,
the 7th and concluding book of
THE EARTH CHRONICLES
now available in hardcover
from William Morrow,
an imprint of HarperCollins Publishers

FOREWORD: THE PAST, THE FUTURE

"When will they return?"

I have been asked this question countless times by people who have read my books, the "they" being the Anunnaki—the extraterrestrials who had come to Earth from their planet Nibiru and were revered in antiquity as gods. Will it be when Nibiru in its elongated orbit returns to our vicinity, and what will happen then? Will there be darkness at noon and the Earth shall shatter? Will it be Peace on Earth, or Armageddon? A Millennium of trouble and tribulations, or a messianic Second Coming? Will it happen in 2012, or later, or not at all?

These are profound questions that combine people's deepest hopes and anxieties with religious beliefs and expectations, questions compounded by current events: Wars in lands where the entwined affairs of gods and men began; the threats of nuclear holocausts; the alarming ferocity of natural disasters. They are questions that I dared not answer all these years—but now are questions the answers to which cannot—must not—be delayed.

Questions about the Return, it ought to be realized, are not new; they have inexorably been linked in the past—as they are today—to the expectation and the apprehension of the Day of the Lord, the End of Days, Armageddon. Four millennia ago, the Near East witnessed a god and his son promising Heaven on Earth. More than three millennia ago, king and people in Egypt yearned for a messianic time. Two millennia ago, the people of Judea wondered whether the Messiah had appeared, and we are still seized with the mysteries of those events. Are prophecies coming true?

We shall deal with the puzzling answers that were given, solve ancient enigmas, decipher the origin and meaning of symbols—the Cross, the Fishes, the Chalice. We shall describe the role of space-related sites in historic events, and show why Past, Present, and Future converge in Jerusalem, the place of the "Bond Heaven-Earth." And we shall ponder why it is that our current twenty-first century A.D. is so similar to the twenty-first century B.C. Is history repeating itself? Is it destined to repeat itself? Is it all guided by a Messianic Clock? Is the time at hand?

More than two millennia ago, Daniel of Old Testament fame repeatedly asked the angels: *When?* When will be the End of Days,

The End of Days

the End of Time? More than three centuries ago the famed Sir Isaac Newton, who elucidated the secrets of celestial motions, composed treatises on the Old Testament's Book of Daniel and the New Testament's Book of Revelation; his recently found handwritten calculations concerning the End of Days will be analyzed, along with more recent predictions of The End.

Both the Hebrew Bible and the New Testament asserted that the secrets of the Future are imbedded in the Past, that the destiny of Earth is connected to the Heavens, that the affairs and fate of Mankind are linked to those of God and gods. In dealing with what is yet to happen, we cross over from history to prophecy; one cannot be understood without the other, and we shall report them both.

It is with that as our guide, let us look at what is to come through the lens of what had been. The answers will be certain to surprise.

Zecharia Sitchin
New York, August 2006

CHAPTER I: THE MESSIANIC CLOCK

Wherever one turns, humankind appears seized with Apocalyptic trepidation, Messianic fervor, and End of Time anxiety.

Religious fanaticism manifests itself in wars, rebellions, and the slaughter of "infidels." Armies amassed by Kings of the West are warring with Kings of the East. A Clash of Civilizations shakes the foundations of traditional ways of life. Carnage engulfs cities and towns; the high and the mighty seek safety behind protective walls. Natural calamities and ever-intensifying catastrophies leave people wondering: Has Mankind sinned, is it witnessing Divine Wrath, is it due for another annihilating Deluge? Is this the Apocalypse? Can there be—will there be—Salvation? Are Messianic times afoot?

The time—the twenty-first century A.D., or was it the twenty-first century B.C.?

The correct answer is Yes and Yes, both in our own time as well as in those ancient times. It is the condition of the present time, as well as at a time more than four millennia ago; and the amazing similarity is due to events in the middle-time in-between—the period associated with the messianic fervor at the time of Jesus.

Those three cataclysmic periods for Mankind and its planet—two in the recorded past (circa 2100 B.C. and when B.C. changed to A.D.), one in the nearing future—are interconnected; one has led to the other, one can be understood only by understanding the other. The Present stems from the Past, the Past is the Future. Essential to all three is **Messianic Expectation**; and linking all three is **Prophecy**.

How the present time of troubles and tribulations will end—what the Future portends—requires entering the realm of Prophecy. Ours will not be a melange of newfound predictions whose main magnet is fear of doom and End, but a reliance upon unique ancient records that documented the Past, predicted the Future, and recorded previous Messianic expectations—prophesying the future in antiquity, and, one believes, the Future that is to come.

In all three apocalyptic instances—the two that had occurred, the one that is about to happen—the physical and spiritual relationship between Heaven and Earth was and remains pivotal for the events. The physical aspects were expressed by the existence

The End of Days

on Earth of actual sites that linked Earth with the heavens—sites that were deemed crucial, that were focuses of the events; the spiritual aspects have been expressed in what we call Religion. In all three instances, a changed relationship between Man and God was central; except that when, circa 2100 B.C., Mankind faced the first of these three epochal upheavals, the relationship was between men and *gods*, in the plural. Whether that relationship has really changed, the reader will soon discover.

The story of the gods, the **Anunnaki** ("Those who from heaven to Earth came"), as the Sumerians called them, begins with their coming to Earth from **Nibiru** in need of gold. The story of their planet was told in antiquity in the *Epic of Creation*, a long text on seven tablets; it is usually considered to be an allegorical myth, the product of primitive minds that spoke of planets as living gods combating each other. But as I have shown in my book *The Twelfth Planet*, the ancient text is in fact a sophisticated cosmogony that tells how a stray planet, passing by our solar system, collided with a planet called Tiamat; the collision resulted in the creation of Earth and its Moon, of the Asteroid Belt and comets, and in the capture of the invader itself in a great elliptical orbit that takes about 3,600 Earth years to complete (Fig. 1 in the hardcover edition).

It was, Sumerian texts tell, 120 such orbits—432,000 Earth years—prior to the Deluge (the "Great flood") that the Anunnaki came to Earth. How and why they came, their first cities in the E.DIN (the biblical Eden), their fashioning of the Adam and the reasons for it, and the events of the catastrophic Deluge—have all been told in *The Earth Chronicles* series of my books and will not be repeated here. But before we time-travel to the momentous twenty-first century B.C., some pre-Diluvial and post-Diluvial landmark events need to be recalled.

The biblical tale of the Deluge, starting in chapter 6 of Genesis, ascribes its conflicting aspects to a sole deity, Yahweh, who at first is determined to wipe Mankind off the face of the Earth, and then goes out of his way to save it through Noah and the Ark. The earlier Sumerian sources of the tale ascribe the disaffection with Mankind to the god **Enlil** and the counter-effort to save Mankind to the god **Enki**. What the Bible glossed over for the sake of Monotheism was not just the disagreement between the Enlil and Enki, but a rivalry and a conflict between two clans of Anunnaki that dominated the course of subsequent events on Earth.

The Messianic Clock

That conflict between the two and their offspring, and the Earth regions allocated to them after the Deluge, need to be kept in mind to understand all that happened thereafter.

The two were half-brothers, sons of Nibiru's ruler **Anu**; their conflict on Earth had its roots on their home planet, Nibiru. Enki—then called E.A ("He whose home is Water")—was Anu's firstborn son, but not by the official spouse, Antu. When Enlil was born to Anu by Antu—a half-sister of Anu—Enlil became the Legal Heir to Nibiru's throne though he was not the firstborn son. The unavoidable resentment on the part of Enki and his maternal family was exacerbated by the fact that Anu's accession to the throne was problematic to begin with: Having lost out in a succession struggle to a rival named Alalu, he later usurped the throne in a coup-d'etat, forcing Alalu to flee Nibiru for his life. That not only backtracked Ea's resentments to the days of his forebears, but also brought about other challenges to the leadership of Enlil, as told in the epic *Tale of Anzu*. (For the tangled relationships of Nibiru's royal families and the ancestries of Anu and Antu, Enlil and Ea, see *The Lost Book of Enki*).

The key to unlocking the mystery of the gods' succession (and marriage) rules was my realization that these rules also applied to the people chosen by them to serve as their proxies to Mankind. It was the biblical tale of the Patriarch Abraham explaining (*Genesis* 20:12) that he did not lie when he had presented his wife Sarah as his sister: "Indeed, she is my sister, the daughter of my father, but not the daughter of my mother, and she became my wife." Not only was marrying a half-sister from a different mother permitted, but a son by her—in this case Isaac—became the Legal Heir and dynastic successor, rather the Firstborn Ishmael, the son of the handmaiden Hagar. (How such succession rules caused the bitter feud between Ra's divine descendants in Egypt, the half-brothers Osiris and Seth who married the half-sisters Isis and Nephtys, is explained in *The Wars of Gods and Men*).

Though those succession rules appear complex, they were based on what those who write about royal dynasties call "bloodlines"— what we now should recognize as sophisticated DNA genealogies that also distinguished between general DNA inherited from the parents as well as the mtDNA that is inherited by females only from the mother. The complex yet basic rule was this: Dynastic lines continue through the male line; the Firstborn son is next in succession; a half-sister could be taken as wife *if she had a different mother*; and if a son by such a half-sister is later born, that son—though not Firstborn—became the Legal Heir and the dynastic successor.

The rivalry between the two half-brothers Ea/Enki and Enlil in matters of the throne was complicated by personal rivalry in matters of the heart. They both coveted their half-sister **Ninmah**, whose mother was yet another concubine of Anu. She was Ea's true love, but he was not permitted to marry her. Enlil then took over and had a son by her—**Ninurta**. Though born without wedlock, the succesion rules made Ninurta Enlil's uncontested heir, being both his Firstborn son and one born by a royal half-sister.

Ea, as related in *The Earth Chronicles* books, was the leader of the first group of fifty Anunnaki to come to Earth to obtain the gold needed to protect Nibiru's dwindling atmosphere. When the initial plans failed, his half-brother Enlil was sent to Earth with more Anunnaki for an expanded Mission Earth. If that was not enough to create a hostile atmosphere, Ninmah too arrived on Earth to serve as chief medical officer . . .

A long text known as the *Atrahasis Epic* begins the story of gods and men on Earth with a visit by Anu to Earth to settle once and for all (he hoped) the rivalry between his two sons that was ruining the vital mission; he even offered to stay on Earth and let one of the half-brothers assume the regency on Nibiru. With that in mind, the ancient text tells us, lots were drawn to determine who shall stay on Earth and who shall sit on Nibiru's throne:

> The gods clasped hands together,
> had cast lots and had divided:
> Anu went up [back] to heaven,
> [For Enlil] the Earth was made subject;
> The seas, enclosed as with a loop,
> to Enki the prince were given.

The result of drawing lots, then, was that Anu returned to Nibiru as its king. Ea, given dominion over the seas and waters (in later times, "Poseidon" to the Greeks and "Neptune" to the Romans), was granted the epithet EN.KI ("Lord of Earth") to soothe his feelings; but it was EN.LIL ("Lord of the Command") who was put in overall charge: "To him the Earth was made subject." Resentful or not, Ea/Enki could not defy the rules of succession or the results of the drawing of lots; and so the resentment, the anger at justice denied, and a consuming determination to avenge injustices to his father and forefathers and thus to himself, led Enki's son **Marduk** to take up the fight.

Several texts describe how the Anunnaki set up their settlemernts in the E.DIN (The post-Diluvial Sumer), each with a specific function, and all laid out in accordance with a master plan. The crucial space connection—the ability to constantly stay in communication with the home planet and with the shuttlecraft and spacecraft—was maintained from Enlil's command post in **Nippur**, the heart of which was a dimly lit chamber called the DUR.AN.KI, "The Bond Heaven-Earth." Another vital facility was a spaceport, located at Sippar ("Bird City"). Nippur lay at the center of concentric circles at which the other "cities of the gods" were located; all together they shaped out, for an arriving spacecraft, a landing corridor whose focal point was the Near East's most visible topographic feature—the twin peaks of Mount Ararat (Fig. 2 in the hardcover edition).

And then the Deluge "swept over the earth," obliterated all the cities of the gods with their Mission Control Center and Spaceport, and buried the Edin under millions of tons of mud and silt. Everything had to be done all over again—but much could no longer be the same. First and foremost, it was necessary to create a new spaceport facility, with a new Mission Control Center and new beacon-sites for a Landing Corridor. The new landing path was anchored again on the prominent twin peaks of Ararat; the other components were all new: The actual spaceport in the Sinai Peninsula, on the 30th parallel north; artificial twin peaks as beacon-sites, the Giza pyramids; and a new Mission Control Center at a place called Jerusalem (Fig. 3 in the hardcover edition). It was a layout that played a crucial role in post-Diluvial events.

The Deluge was a watershed (both literally and figuratively) in the affairs of both gods and men, and in the relationship between the two: The Earthlings, who were fashioned to serve and work for the gods, were henceforth treated as junior partners on a devastated planet.

The new relationship between men and gods was formulated, sanctified and codified when Mankind was granted its first high civilization, in Mesopotamia, circa 3800 B.C. The momentous event followed a state visit to Earth by Anu, not just as Nibiru's ruler but also as the head of the pantheon, on Earth, of the ancient gods.

The End of Days

Another (and probably the main) reason for his visit was the establishment and affirmation of peace among the gods themselves—a live-and-let-live arrangement by dividing the lands of the Old World among the two principal Anunnaki clans—that of Enlil and that of Enki; for the new post-Diluvial circumstances and the new location of the space facilities required a new territorial division among the gods.

It was a division that was reflected in the biblical Table of Nations (*Genesis*, chapter 10), in which the spread of Mankind, emanating from the three sons of Noah, was recorded by nationality and geography: Asia to the nations/lands of Shem, Europe to the descendants of Japhet, Africa to the nation/lands of Ham. The historical records show that the parallel division among the gods allotted the first two to the Enlilites, the third one to Enki and his sons. The connecting Sinai peninsula, where the vital post-Diluvial spaceport was located, was set aside as a neutral Sacred Region.

While the Bible simply listed the lands and nations according to their Noahite division, the earlier Sumerian texts recorded the fact that the division was a deliberate act, the result of deliberations by the leadership of the Anunnaki. A text known as the *Epic of Etana* tells us that

> The great Anunnaki who decree the fates
> sat exchanging their counsels regarding the Earth.
> They created the four regions,
> set up the settlements.

In the First Region, the lands between the two rivers Euphrates and Tigris (Mesopotamia), Man's first known high civilization, that of Sumer, was established. Where the pre-Diluvial cities of the gods had been, Cities of Man arose, each with its sacred precinct where a deity resided in his or her ziggurat—Enlil in Nippur, Ninmah in Shuruppak, Ninurta in Lagash, **Nannar/Sin** in Ur, **Inanna/Ishtar** in Uruk, **Utu/Shamash** in Sippar, and so on. In each such urban center an EN.SI, a "Righteous Shepherd"—initially a chosen demigod—was selected to govern the people in behalf of the gods; his main assignment was to promulgate codes of justice and morality. In the sacred precinct, a priesthood overseen by a high priest served the god and his spouse, supervised the holiday celebrations, and handled the rites of offerings, sacrifices and prayers to the gods. Art and sculpture, music and dance, poetry

The Messianic Clock

and hymns, and above all writing and recordkeeping flourished in the temples and extended to the royal palace.

From time to time one of those cities was selected to serve as the land's capital; there the ruler was king, LU.GAL ("Great man"). Initially and for a long time thereafter this person, the most powerful man in the land, served as both king and high priest. He was carefully chosen, for his role and authority, and all the physical symbols of Kingship, were deemed to have come to Earth directly from Heaven, from Anu on Nibiru. A Sumerian text dealing with the subject stated that before the symbols of Kingship (tiara/crown and scepter) and of Righteousness (the shepherd's staff) were granted to an earthly king, they "lay deposited before Anu in heaven." Indeed, the Sumerian word for Kingship was *Anuship*.

This aspect of "Kingship" as the essence of civilization, just behavior and a moral code for Mankind, was explicitly expressed in the statement, in the Sumerian King Lists, that after the Deluge *"Kingship was brought down from Heaven."* It is a profound statement that must be borne in mind as we progress in this book to the messianic expectations—in the words of the New Testament, for the **Return of the "Kingship of Heaven" to Earth.**

Circa 3100 B.C. a similar yet not identical civilization was established in the Second Region in Africa, that of the river Nile (Nubia and Egypt). Its history was not as harmonious as that among the Enlilites, for rivalry and contention continued among Enki's six sons, to whom not cities but whole land domains were allocated. Paramount was an ongoing conflict between Enki's firstborn **Marduk** (*Ra* in Egypt) and **Ningishzidda** (*Thoth* in Egypt), a conflict that led to the exile of Thoth and a band of African followers to the New World (where he became known as *Quetzaloatl*, the Winged Serpent). Marduk/Ra himself was punished and exiled when, opposing the marriage of his young brother Dumuzi to Enlil's granddaughter Inanna/Ishtar, he caused his brother's death. It was as compensation to Inanna/Ishtar that she was granted dominion over the Third Region of civilization, that of the Indus Valley, circa 2900 B.C. It was for good reason that the three civilizations—as was the spaceport in the sacred region—were all centered on the 30th parallel north (Fig. 4 in the hardcover edition).

According to Sumerian texts, the Anunnaki established Kingship—civilization and its institutions, as most clearly exemplified in Mesopotamia—as a new order in their relationships with Mankind, with kings/priests serving both as a link and a separator between gods and men. But as one looks back on that seemingly "golden age" in the affairs of gods and men, it becomes evident that the affairs of the gods constantly dominated and determined the affairs of Men and the fate of Mankind. Overshadowing all was the determination of Marduk/Ra to undo the injustice done to his father Ea/Enki, when under the succession rules of the Anunnaki not Enki but Enlil was declared the Legal Heir of their father Anu, the ruler on their home planet Nibiru.

In accord with the sexagesimal ("base sixty") mathematical system that the gods granted the Sumerians, the twelve great gods of the Sumerian pantheon were given numerical ranks in which Anu held the supreme Rank of Sixty; the Rank of Fifty was granted to Enlil; that of Enki was 40, and so farther down, alternating between male and female deities (Fig. 5 in the hardcover edition). Under the succession rules, Enlil's son Ninurta was in line for the rank of 50 on Earth, while Marduk held a nominal rank of 10; and initially, these two successors-in-waiting were not yet part of the twelve "Olympians."

And so, the long, bitter and relentness struggle by Marduk that began with the Enlil-Enki feud focused later on Marduk's contention with Enil's son Ninurta for the succession to the Rank of Fifty, and then extended to Enlil's granddaughter Inanna/Ishtar whose marriage to Dumuzi, Enki's youngest son, was so opposed by Marduk that it ended with Dumuzi's death. In time Marduk/Ra faced conflicts even with other brothers and half-brothers of his, in addition to the conflict with Thoth that we have already mentioned—principally with Enki's son Nergal who married a granddaughter of Enlil named Ereshkigal.

In the course of these struggles, the conflicts at times flared up to full-fledged wars between the two divine clans; some of those wars are called "The Pyramid Wars" in my book *The Wars of Gods and Men*. In one notable instance the fighting led to the burying alive of Marduk inside the Great Pyramid; in another, it led to its capture by Ninurta. Marduk was also exiled, more than once—both as punishment and as a self-imposed absence. His persistent efforts to attain the status to which he believed he was entitled included the event recorded in the Bible as the Tower of Babel incident; but in the end, after numerous frustrations, success came only when Earth and Heaven were aligned with the **Messianic Clock**.

Indeed, the first cataclysmic set of events, in the twenty-first century B.C., and the Messianic expectations that accompanied it, is principally the story of Marduk; it also brought to center stage his son **Nabu**—a deity, the son of a god, but whose mother was an Earthling.

Throughout the history of Sumer that spanned almost two thousand years, its royal capital shifted—from the first one, Kish (Ninurta's first city) to Uruk (the city that Anu granted to Inanna/Ishtar) to Ur (Sin's seat and center of worship); then to others and then back to the initial ones; and finally, for the third time, back to Ur. But at all times Enlil's city Nippur, his "cult center" as scholars are wont to call it, remained the religious center of Sumer and the Sumerian people; it was there that the annual cycle of worshipping the gods was determined.

The twelve "Olympians" of the Sumerian pantheon, each with his or her celestial counterpart among the twelve members of the Solar System (Sun, Moon and ten planets, including Nibiru), were also honored with one month each in the annual cycle of a twelve-month year. The Sumerian term for "month," EZEN, actually meant holiday, festival; and each such month was devoted to celebrating the worship-festival of one of the twelve supreme gods. It was the need to determine the exact time when each such month began and ended (and not in order to enable peasants to know when to sow or harvest, as schoolbooks explain) that led to the introduction of *Mankind's first calendar* in **3760** B.C. It is known as the **Calendar of Nippur** because it was the task of its priests to determine the calendar's intricate timetable and to announce, for the whole land, the time of the religious festivals. That calendar is still in use to this day as the Jewish religious calendar which, in A.D. 2006, numbered the year as 5766.

In pre-Diluvial times Nippur served as Mission Control Center, Enlil's command post where he set up the DUR.AN.KI, the "Bond Heaven-Earth" for the communications with the home planet Nibiru and with the spacecraft connecting them. (After the Deluge, these functions were relocated to a place later known as Jerusalem). Its central position, equidistant from the other functional centers in the E.DIN (*see* Fig. 2), was also deemed to be equidistant from the "four corners of the Earth" and gave it the nickname *"Navel of the Earth."* A hymn to Enlil referred to Nippur and its functions thus:

The End of Days

Enlil,
When you marked off divine settlements on Earth,
Nippur you set up as your very own city...
You founded the Dur-An-Ki
In the center of the four corners of the Earth.

(The term "the Four Corners of the Earth" is also found in the Bible; and when Jerusalem replaced Nippur as Mission Control Center after the Deluge, it too was nicknamed the Navel of the Earth).

In Sumerian the term for the four regions of the Earth was UB, but it also is found as AN.UB—the heavenly, the *celestial* four "corners"—in this case an astronomical term connected with the calendar. It is taken to refer to the four points in the Earth-Sun annual cycle that we nowadays call the Summer Solstice, the Winter Solstice, and the two crossing of the equator—once as the Spring Equinox and then as the Autumnal Equinox. In the Calendar of Nippur, the year began on the day of the Spring Equinox and it has so remained in the ensuing calendars of the ancient Near East. That determined the time of the most important festival of the year—the New Year festival, an event that lasted ten days during which detailed and canonized rituals had to be followed.

Determining calendrical time by Heliacal Rising entailed the observation of the skies at dawn, when the sun just begins to rise on the eastern horizon but the skies are still dark enough to show the stars in the background. The day of the equinox having been determined by the fact that on it daylight and nighttime were precisely equal, the position of the sun at heliacal rising was then marked by the erection of a stone pillar to guide future observations—a procedure that was followed, for example, later on at Stonehenge in Britain; and, as at Stonehenge, long term observations revealed that the group of stars ("constellation") in the background has not remained the same (Fig. 6 in the hardcover edition); there, the alignment stone called the "Heel Stone" that points to sunrise on solstice day nowadays pointed originally to sunrise circa 2000 B.C.

The phenomenon, called Precession of the Equinoxes or just Precession, results from the fact that as the Earth completes one annual orbit around the Sun, it does not return to the same exact celestial spot. There is a slight, very slight retardation; it amounts to one degree (out of 360 in the circle) in 72 years. It was Enki who first grouped the stars observable from Earth into "constellations," and divided the heavens in which the Earth circled the sun into twelve parts—what has since been called the Zodiacal Circle of constellations (Fig. 7 in the hardcover edition). Since each twelfth part of the circle occupied 30 degrees of the celestial arc, the retardation or Precessional shift from one Zodiacal House to another lasted (mathematically) 2160 years (72 x 30), and a complete zodiacal cycle lasted 25,920 years (2160 x 12). The approximate dates of the **Zodiacal Ages**—following the equal twelve-part division and not actual astronomical observations—have been added here for the readere's guidance.

That this was the achievement from a time preceding Mankind's civilizations is attested by the fact that a zodiacal calendar was applied to Enki's first stays on Earth (when the first two zodiacal houses were named in his honor); that this was not the achievement of a Greek astronomer (Hipparchus) in the third century B.C. (as most textbooks still suggest) is attested by the fact that the twelve zodiacal houses were known to the Sumerians millennia earlier by names (Fig. 8 in the hardcover edition) and depictions (Fig. 9 in the hardcover edition) that we use to this day.

In *When Time Began* the calendrical timetables of gods and men were discussed at length. Having come from Nibiru, whose orbital period, the SAR, meant 3,600 (Earth-) years, that unit was naturally the first calendrical yardstick of the Anunnaki even on the fast-orbiting Earth. Indeed, the texts dealing with their early days on Earth, such as the Sumerian King Lists, designated the periods of this or that leader's time on Earth in terms of Sars. I termed this **Divine Time**. The calendar granted to Mankind, one based on the orbital aspects of the Earth (and its Moon), was named **Earthly Time**. Pointing out that the 2160-year zodiacal shift (less than a year for the Anunnaki) offered them a better ratio—the "golden ratio" of 10:6—between the two extremes; I called this time unit **Celestial Time**.

As Marduk discovered, that Celestial Time was the "clock" by which his destiny was to be determined.

But which was **Mankind's Messianic Clock**, determining its fate and destiny—*Earthly Time*, such as the count of fifty-year Jubilees, a count in centuries, or the Millennium? Was it *Divine Time*, geared to Nibiru's orbit? Or was it—is it—*Celestial Time* that follows the slow rotation of the zodiacal clock?

The quandary, as we shall see, baffled Mankind in antiquity; it still lies at the core of the current Return issue. The question that

The End of Days

is posed has been asked before—by Babylonian and Assyrian star-gazing priests, by biblical Prophets, in the Book of Daniel, in the Revelation of St. John the Divine, by the likes of Sir Isaac Newton, by all of us today.

The answer will be astounding. Let us embark on the painstaking quest.